# PHYSICS
# 100 IDEAS IN 100 WORDS

**A whistle-stop tour
of key concepts**

SCIENCE
MUSEUM

# PHYSICS
# 100
# IDEAS
# IN 100
# WORDS

## A whistle-stop tour of key concepts

David Sang

# Contents

# Introduction

We can think of physics as starting with astrology, several millennia ago. At a time when there was less light pollution to reduce the visibility of the stars in the night sky, people were more familiar with the pattern of the constellations. They knew that certain stars were only visible at particular seasons. They observed that some stars seemed different, changing their positions against the pattern of the fixed stars – these were the planets. The appearance of the Moon changed from night to night. Occasionally, people observed eclipses of the Sun and Moon.

Astrologers learned to record these changing patterns, to measure the positions of the heavenly bodies and to predict events in the night sky. They did this because they wanted to know the future. They attributed significance to their observations – who should be the new leader, when was a good time to go to war, why did a crop fail?

Many of the basic elements of physics were there in the work of the astrologers: careful observation, measurement, recording, calculation, prediction. Their findings were useful to the first modern astronomers, two or three millennia later. However, while it's a natural human impulse to look for meaning in what we see around us, astrology failed because no one was able to produce reliable evidence that the movement of the planets through the constellations has a direct impact on human lives down here on Earth.

Why does the pattern of the constellations move across the night sky, and change with the seasons? Why do the planets behave differently? Are there laws governing this? Think closer to home. Drop a ball and watch it bounce. Drop it again and it does the same thing. It never bounces higher than the point from which you dropped it. It's predictable; is it obeying some hidden law? Shout into a cave and you'll hear an echo. Why is the echo like looking in a mirror? Why does a compass needle point north–south, but only approximately?

Physics can answer these questions and has gone far beyond them. At the end of the 19th century, there was a debate among physicists – was there much more to be discovered? Many thought that the basic laws of physics were well established and that all that remained was to apply them in different situations. But 20th-century discoveries proved them wrong, opening up the strange worlds of particle physics, quantum mechanics and relativity.

In the 1920s, Cecilia Payne, a young British physicist, studied starlight to discover the composition of stars. An older generation of astronomers assumed that stars were made of a similar mix of elements to the planets, but Cecilia showed that they were wrong. The stars are made largely of hydrogen and helium, the first elements created in the big bang. Today we know that many elements are made in dying stars as they explode into space. This material condenses to form new stars and planets, and eventually some of it forms the bodies of humans and other living things. We are stardust. What would those ancient astrologers have made of that?

# Matter

Physics is a way of looking at the world and trying to make sense of it. Books, furniture, people, houses, trees, water, the Sun, stars at night – the things surrounding us may all appear to be very different from each other, but physics says that they are all made of matter.

A defining approach in physics is to seek a single explanation for varied phenomena. Two millennia ago, people imagined that there were four different elements – earth, water, air, fire – and believed that they had no way of knowing what astronomical objects such as the Moon and Sun were made of. Two centuries ago, people would have been surprised to learn that living things are made of just the same stuff as rocks, water and stars.

Brownian motion is the random motion of tiny grains of material suspended in a liquid or a gas (a fluid), caused by the grains being struck randomly by the particles that make up the fluid. It is strong evidence for the idea that all matter is made up of particles, too small to be seen even with the most powerful microscope, and in constant motion. In a gas, these particles are free to move within the volume enclosed by its container.
In 1905 Albert Einstein used Brownian motion to calculate the number of particles in a given **mass** of gas.

**1**

**WHY IT MATTERS**
Describes the macroscopic evidence for the microscopic nature of matter

**KEY THINKERS**
Robert Brown
(1773–1858)
Albert Einstein
(1879–1955)
Jean Perrin
(1870–1942)

**WHAT COMES NEXT**
The kinetic model of matter

**SEE ALSO**
Mass and volume, p.12
Structures of materials, p.16

# Brownian motion

Looking through his microscope at pollen grains in water in 1827, Scottish botanist Robert Brown noticed tiny scraps that seemed to be constantly moving around. Their jittery movements seemed random – could this be because they were fragments of living material? But when Brown looked at fragments of rock in water he saw the same thing. Dead or alive, scraps of stuff showed the same jittery motion.

Now we know that the tiny solid scraps he was examining were moving because they were being constantly bombarded by the molecules of the water in which they were suspended. Water molecules are far too small to be seen using a light microscope, but their effects on bigger chunks of material could be seen. You can see the same effect by looking at tiny grains of dust in air – air molecules, like water molecules, are in constant motion.

# 2

# The kinetic model of matter

Physicists picture matter as being made of particles. These may be atoms, molecules or **ions** – it's not important for the particle **model** of matter.

If you could cool the air in a room to about –200°C it would condense into a liquid and form a layer just 2–3mm deep on the floor.

This tells us that there's a lot of empty space in a gas such as air – the particles of which it is made are far apart, with a vacuum in between. That's why it's much easier to walk through air than move through water, where the particles are close together.

It also tells us that the particles of matter attract one another. Cooling a gas slows down its particles. If they move slowly the attractive forces between them cause them to stick together, forming first a liquid and then a solid.

STATES OF MATTER

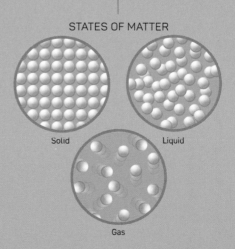

Solid

Liquid

Gas

The kinetic **model** of matter explains the three states of matter – solid, liquid and gas – by describing the arrangement and motion of the particles of which matter is made.
Kinetic means moving.
In a solid, the particles are packed closely and vibrate about their fixed positions. In a liquid, the particles are packed closely but can slide past one another, while in a gas, they are spread widely apart and can move around freely.
This explains why a solid has a fixed size and shape, why a liquid can flow and why a gas fills its container.

"So many of the properties of matter, especially when in the gaseous form, can be deduced from the hypothesis that their minute parts are in rapid motion, the velocity increasing with the temperature, that the precise nature of this motion becomes a subject of rational curiosity."
**James Clerk Maxwell**, physicist

# 3

# Mass and volume

A gas such as air is much lighter than a solid such as gold. This is due partly to the empty space between air particles but also because gold particles have greater **mass** than air particles. Gold has greater **density** than air – the mass of its particles is compressed into a much smaller volume.

Gold and silver are similar metals. Their atoms are almost exactly the same size, but atoms of gold have 1.8 times the mass of silver atoms, so gold is 1.8 times more dense than silver.

## WHY IT MATTERS
**Mass** and volume are two fundamental properties of matter

## KEY THINKERS
Thales of Miletus (624–548BCE)
Democritus (460–370BCE)
René Descartes (1596–1650)

## WHAT COMES NEXT
Understanding **density** in terms of particle theory

## SEE ALSO
The kinetic model of matter, p.10

In
**100**
words

**Mass**, measured in kilograms (kg), and volume, measured in cubic metres ($m^3$), are two fundamental properties of matter.
The **density** of a material arises from the mass of its particles and how close they are to each other. If matter is compressed, its volume is reduced. This increases its density. However, its volume cannot be reduced to zero, no matter how great the pressure.
When air is cooled until it becomes a liquid at around -200°C, it shrinks to about one-thousandth of its starting volume, showing that 99.9% of the volume of air is the vacuum between its molecules.

## 4

The pressure of a fluid (a gas or liquid) on the walls of its container is the force exerted on each square metre of the surface. It is caused by the particles of the fluid colliding with the walls of the container. The pressure increases if the particles collide more frequently with the walls.

Boyle's law relates the pressure of a gas to its volume. Decreasing the volume of a gas causes the particles to collide more frequently with the container walls and so the pressure increases. Halving the volume of a gas gives double the pressure, and vice versa.

**WHY IT MATTERS**
The law relates the volume of a gas to its pressure

**KEY THINKERS**
Robert Boyle
(1627–1691)
Katherine Boyle,
Viscountess Ranelagh
(1615–1691)
Robert Hooke
(1635–1703)
Edmé Mariotte
(1620–1684)

**WHAT COMES NEXT**
Thermodynamics – the laws that govern how matter behaves when heated or cooled

**SEE ALSO**
The kinetic model of matter, p.10
Charles's law, p.14

# Boyle's law

Air particles at room temperature move faster than the speed of sound – at close to 400 m/s. When they collide with a solid surface they bounce off, and each collision creates a tiny force on the surface. Because vast numbers of particles are colliding each second, there is a sizeable force on each square metre of the surface. This is the origin of the pressure of a gas.

The invention of the vacuum pump made experiments on gases possible. Using such a pump, Robert Boyle was able to show how the pressure of a gas changes as it is compressed to a smaller volume. This was the first of the "gas laws", which became important for understanding how gases behave when they are compressed or heated in an engine or piston.

# 5

# Charles's law

**WHY IT MATTERS**
The law suggests that
there is a minimum
temperature that
cannot be passed

**KEY THINKERS**
John Dalton
(1766–1844)
Jacques Charles
(1746–1823)
William Thomson
(1824–1907)

**WHAT COMES NEXT**
Understanding the
physical properties of
gases made steam
engines possible

**SEE ALSO**
Kinetic energy, p.84
Absolute zero, p.96

French physicist Jacques Charles investigated how the volume of a gas changes when it is cooled. To cool a gas the movement of its particles must be slowed down – their **kinetic energy** (see page 84) must be reduced. The volume of the gas decreases provided that its pressure remains the same. Eventually the attractive forces between the molecules cause them to stick together and they form a liquid.

In the early 20th century physicists discovered how to cool gases sufficiently to turn them into liquids. Helium was the most difficult – it must be cooled to -269°C. Helium is difficult to liquify because the forces between its particles are weak so that they don't start sticking together until they are moving very slowly.

John Dalton demonstrated that all gases show the same pattern: as the temperature is reduced the volume of the gas decreases by the same amount for every 10-degree drop in temperature.

Cooling a gas causes its particles to move more slowly and the external pressure pushing inwards causes its volume to decrease. If it didn't liquify, its volume would reduce to zero at a temperature of about -273°C, known as **absolute zero** (see page 96). This is true for all gases. Charles's law says that the volume of a gas is proportional to its absolute temperature. Doubling the temperature of a gas causes its particles to move faster so that they push the walls of the container further apart. The volume doubles, provided that the pressure doesn't change.

6

# Structures of materials

**WHY IT MATTERS**
Understanding how the structure of a material gives rise to its properties enables new materials to be designed

**KEY THINKERS**
Robert Hooke
(1635–1703)
Sophie Germain
(1776–1831)
Katharine Blodgett
(1898–1979)
Mildred Dresselhaus
(1930–2017)
Harry Kroto
(1939–2016)
Andre Geim
(1958–)

**WHAT COMES NEXT**
Metals, ceramics, polymers, semiconductors – all can be manipulated to give desired properties

**SEE ALSO**
Atomic structure, p.105

We can think of the particles that make up matter as tiny spheres. The attractive forces between neighbouring particles form **bonds** that hold the particles together in a solid.

There are many ways spheres can bond. Carbon atoms make an interesting example. Diamonds are made entirely of carbon atoms – each atom is bonded to four others in a three-dimensional array, making diamond an exceedingly hard material.

Graphite, the black material used in soft pencils, is also a form of carbon. But its atoms are bonded to just three neighbours to form a flat sheet like a honeycomb. There is only weak attraction between sheets, so graphite is a weak material, soft and perfect for drawing or writing with a pencil.

Two other forms of carbon are known. In soot, carbon atoms bond in groups of 60 to form hollow spheres known as buckminsterfullerene. Graphene consists of a single sheet of carbon atoms, similar to a single layer of graphite.

The attractive forces between particles create **bonds** between them. Particles can bond in different ways, giving a material its underlying structure.

The physical properties of a material depend on its structure and the strength of the bonds between its particles. For example, a material whose particles are strongly bonded is likely to be stiff (difficult to stretch) and strong (hard to break). Metals are malleable (they can be hammered into shape) because their particles are less strongly bonded to each other and can slide past one another, allowing the metal to deform when a force is applied.

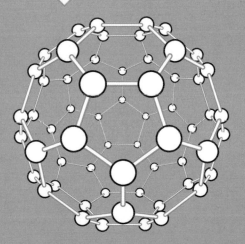

In this fullerene molecule, each of 60 carbon atoms is bonded to three others

# 7

# Chemistry and biochemistry

In physics we tend to think of the particles that make up matter as simple, identical spheres. Chemists study the many ways different atoms can **bond** to make different substances with complex molecular shapes. Biochemists look at the complex molecules that are produced by living things as a result of 4 billion years of evolution.

**WHY IT MATTERS**
Understanding chemistry and biochemistry is fundamental to industrial processes

**KEY THINKERS**
John Dalton
(1766–1844)
Justus von Liebig
(1803–1873)
Niels Bohr
(1885–1962)
Linus Pauling
(1901–1994)
Eva Nogales
(1965–)

**WHAT COMES NEXT**
Millions of new materials, including medicines

**SEE ALSO**
The periodic table, p.108

In **100** words

The atoms of each chemical **element** are different (see page 108). Atoms of different elements can **bond** in many ways. Chemistry studies how atoms bond to form different substances and uses these ideas to explain the properties of these materials. Biochemistry studies the substances formed by living organisms, seeking to explain how they are produced and used, and how they are the product of evolution. Both chemists and biochemists have created a myriad of new substances not found in nature. The laws of physics underlie the ways in which these substances form and interact with each other.

# Forces and Motion

Our lives are full of movement, from the natural world – people, birds, water, the wind – to human inventions such as cars and aircraft. Until the 16th century, the way things move was taken to be "natural" – a stone falls because its natural place is on the ground. Smoke rises because it has a natural property called "levity" that makes it move upwards. The Moon goes round the Earth because that's what celestial bodies do.

However, this changed when scientists – particularly Italian astronomer Galileo Galilei and English physicist Isaac Newton – set about understanding what were the true "laws" that govern the motion of objects, both here on the surface of the Earth and beyond, in space. Although these laws are now well established, they are not immediately obvious.

# 8

# Mass and inertia

The Voyager 1 spacecraft was launched in 1977 and is now approaching the edge of the solar system, more than 20 billion kilometres from the Sun. It has long since used all of the fuel in its rocket engines, but it has no need for fuel because, in the vacuum of space, there is no force such as air resistance to slow it down. It will keep travelling at a steady speed in a straight line forever – or until it is captured by the **gravitational field** of some distant star or planet (see page 74).

**WHY IT MATTERS**
Describes how objects tend to move

**KEY THINKERS**
Aristotle (384–322BCE)
Lucretius (99–55BCE)
Ibn Sina (Avicenna)
(980–1037)

**WHAT COMES NEXT**
How forces affect motion

**SEE ALSO**
Motion under gravity, p.29
Gravitational fields, p.74

## In 100 words

The tendency of an object to keep moving at a constant speed in a straight line is known as **inertia**. Inertia also means that it is difficult to start an object moving when it is at rest. A force must be applied to start it moving.

It takes less force to start a marble rolling than is needed to start a broken-down car moving along the road. This is because a marble has less **mass** than a car.

Mass, sometimes called inertial mass, and measured in kilograms (kg), is the quantity that tells us about the inertia of an object.

If an object is stationary, it remains stationary unless an external force acts on it. If an object is moving, it will continue to move at a constant speed in a straight line, unless an external force acts on it.

This is known as Newton's first law of motion, although Galileo, who died in the year Newton was born, also had these ideas. Motion "at a constant speed in a straight line" is known as uniform motion.

An external force is a push or pull acting on the object, caused by another object. The force of friction is an example.

**WHY IT MATTERS**
Describes motion in the absence of external forces

**KEY THINKERS**
Galileo Galilei
(1564–1642)
Isaac Newton
(1643–1727)

**WHAT COMES NEXT**
The idea of
**acceleration**

**SEE ALSO**
Space travel, p.148

# Newton's first law of motion

If you roll a ball along the ground, it will come to a halt – it is stopped by the force of friction. Our everyday experience is that, to keep an object moving, we need to provide a force. That's why a car has an engine – if you're driving a car, you keep your foot on the accelerator to provide the force needed to keep it going. However, our everyday experience is deceptive. A car is always pushing against the force of friction and air resistance, and against gravity when it is going uphill.

Galileo thought about how an object would move if there were no forces acting on it – no gravity, no resistive forces. He concluded that, if there were no forces acting on it, it would keep moving at a steady speed in a straight line.

# 10

# Acceleration

**WHY IT MATTERS**
Quantifies the rate of
change of speed

**KEY THINKERS**
Galileo Galilei
(1564–1642)

**WHAT COMES NEXT**
Newton's second law of
motion

**SEE ALSO**
Momentum, p.28

Adverts for cars often declare something like "0 to 60 miles per hour in 8.4 seconds", publicising the **acceleration** the engine can provide. To make a car travel faster, you press harder on the accelerator pedal to provide the extra force needed to change its speed – the greater the force, the greater the car's acceleration.

If you drop a ball it accelerates downwards. It's difficult to see this as it takes a fraction of a second to reach the ground. It's easier to see the effect of the Earth's gravity if you throw a ball upwards. It's moving fastest as it leaves your hand. It slows down (decelerates) as it rises until, at the highest point, it is instantaneously stationary. Then its accelerates as it falls downwards.

"It's not magic! It's physics.
The speed of the turn is what
keeps you upright. It's like a
spinning top."
**Deborah Bull**, dancer

An external force is needed to make an object accelerate – to change its speed. The rate at which its speed changes is its **acceleration.** If an object's speed increases by 1 metre per second every second, we say that its acceleration is 1 metre per second per second, written as 1m/s/s or $1m/s^2$. Objects falling freely near the Earth's surface speed up as they fall. Their acceleration is $9.8\ m/s^2$.

Since uniform motion is constant speed in a straight line, an external force acting on an object can also change the direction in which the object is moving.

# 11

# Newton's second law of motion

Isaac Newton's second law relates the **acceleration** of a **mass** to the force acting on it. He published his theories in Latin, in his book *Principia Mathematica* ("Mathematical Principles"). Latin was a shared language among scientists at the time, but its use limited the audience for Newton's ideas.

Scottish mathematician Mary Somerville produced a translation into English which was very influential, making Newton's ideas available to many more people. While in France, Émilie du Châtelet made a French translation that is still read today. Translating Newton was no easy task as the translators had to have a deep understanding of his ideas. Both Somerville and du Châtelet went on to further Newton's ideas and contribute to popularising other branches of science.

Somerville College at the University of Oxford is named after Mary Somerville.

A large force is needed to give a **mass** a large **acceleration**. The greater the mass, the greater the force needed to produce a given acceleration.
Newton's second law of motion combines these ideas into a single relationship: force = mass × acceleration. In this equation the unit of force is the newton (N) (see Mathematics page 180).
This equation tells us what a force is: it is an effect acting on a mass that causes the mass to accelerate. If no force acts, the speed of the mass will be constant – as stated in Newton's first law.

An object with a greater mass will require a greater force to accelerate. The opposite is true for an object with a smaller mass

Small force
*f*

Big mass
*M*

Big force
*F*

Small mass
*m*

Small acceleration

Large acceleration

# 12

# Newton's third law of motion

Bend down and get hold of your bootlaces. No matter how hard you pull, you can't lift yourself off the ground. To make an object move requires an external force – a force caused by another object.

So how can you get yourself off the ground? Bend your knees and press down hard on the ground – the force you exert is pressing downwards. It's the upward force of the ground that makes you move upwards.

Similarly, at the start of a sprint, you push hard backwards on the ground so that it will push you forwards, giving you a flying start. It can be hard to appreciate this because we think of ourselves as active and the ground as passive.

The Moon orbits the Earth. The gravitational pull of the Earth on the Moon keeps it in its orbit, preventing it from flying off into the depths of space. At the same time, we know that the Moon exerts a gravitational pull on the Earth. This is what causes the tides. These two forces, one on the Moon caused by the Earth and the other on the Earth caused by the Moon, are equal in size and act in opposite directions.

If Object A exerts a force on Object B, Object B exerts a force on Object A. These two forces are equal in size and opposite in direction. This is Newton's third law of motion.

The two forces must be of the same type. For example, both must be gravitational (like the forces between the Earth and the Moon) or electrical or magnetic.

A pair of forces as described by Newton's third law are sometimes referred to as "action" and "reaction", but this can be misleading. The two forces act at the same time; one isn't caused by the other.

# 13

# Momentum

Imagine you are an astronaut carrying out repairs on the outside of your space station. Your tether breaks and you find yourself floating slowly away from the station. How can you get back?

You need a force acting on you to push you in the right direction. Throw your spanner as hard as you can, away from the station. By exerting a force on the spanner you produce an equal and opposite force acting on yourself. This will start you moving towards the space station. You may only move slowly, but Newton's first law tells you that you will travel with uniform motion until you collide with the station. You will never see the spanner again.

**WHY IT MATTERS**
Enables us to calculate how objects will move after collisions and explosions

**KEY THINKERS**
René Descartes
(1596–1650)
Galileo Galilei
(1564–1642)
Christiaan Huygens
(1629–1695)
Gottfried Leibniz
(1646–1716)

**WHAT COMES NEXT**
Computer modelling of complex motions

**SEE ALSO**
Newton's third law of motion, p.26

## In 100 words

Newton expressed his laws of motion in terms of **momentum,** calculated by multiplying the **mass** of an object by its **velocity** (its speed in a particular direction): momentum = mass × velocity.
Momentum is a **conserved quantity** – when two objects interact, there is the same total momentum after the interaction as there was before. If one loses momentum, the other must gain an equal amount.
This is a consequence of Newton's third law. The two objects are acted on by equal and opposite forces for the same time, so the effects on each must be equal and opposite.

Weight is the force of the Earth's gravity acting on a **mass**. Since weight is a force it is measured in newtons (N). Close to the Earth's surface each kilogram of mass has a weight of about 9.8N acting downwards, towards the centre of the Earth. The quantity 9.8N/kg is known as the **gravitational field** strength (g). The value of g varies across the Earth's surface; it decreases as you go up a mountain, getting further from the centre of the Earth.

An object falling freely under the effect of gravity will accelerate downwards with an **acceleration** of $9.8\text{m/s}^2$.

**WHY IT MATTERS**
Explains how objects move under gravity

**KEY THINKERS**
Johannes Kepler (1571–1630)
Isaac Newton (1643–1727)

**WHAT COMES NEXT**
Understanding the motion of the planets and other astronomical objects

**SEE ALSO**
Planets, p.146

# Motion under gravity

If you run off the top of a cliff like a cartoon character, you will follow a curved path through the air until you reach the ground below. You'll travel at a steady speed horizontally because there is no horizontal force acting on you. At the same time, you'll **accelerate** vertically downwards, pulled by your weight (the force of the Earth's gravity on you). The shape of this curve is called a parabola. A stone thrown through the air will follow the same shape of path.

# 15

# Pairs of forces

Why don't we fall through the floor? Our feet press down on the floor. This force causes the floor to be very slightly depressed – the effect is obvious if you stand on a trampoline. Newton's third law tells us that there will be an equal and opposite upward force on our feet, which will be enough to support us unless the floor is very weak.

As a jet aircraft flies, hot gases are pushed out of its engines at high speed. The consequence is a forward force on the aircraft, keeping it moving through the air. This force has to be sufficient to counteract the drag of the air, which tends to slow it down. At the same time, the aircraft's wings are angled slightly upwards. As it moves through the air, the wings push air downwards. The result is an upward force on the wings, called lift, ensuring that the aircraft's weight doesn't cause it to fall from the sky.

"My brother and I became seriously interested in the problem of human flight in 1899 ... We knew that men had by common consent adopted human flight as the standard of impossibility. When a man said, 'It can't be done; a man might as well try to fly,' he was understood as expressing the final limit of impossibility."

**Wilbur Wright**, aviation pioneer

## In
## 100
### words

The motion of an object can only change if an external force acts on it. To understand how an object's motion changes we need to identify the external forces acting on it. Newton's third law tells us that, to do this, we must find the equal and opposite forces it exerts on other objects. Humans and other living organisms have evolved to control the forces they exert on other objects (including the ground, air, water and so on) in order to move around. Similarly, an understanding of how forces arise enables engineers and designers to create all kinds of structures.

# 16

# Space-time

Newton's laws came about because he was able to picture how objects would move and interact if they were free from the earthly encumbrances of friction, weight and so on. However, two centuries later, German-born physicist Albert Einstein developed his two theories of relativity, which represent a complete rethink of how physicists see the world.

Newton imagined that we live in a "clockwork universe". We can adopt a godlike point of view, with a ticking clock that enables us to determine the order of events that we observe.

Einstein realised that this is impossible. Suppose you observe two stars exploding, A before B. An observer at a different point in the Universe might see B explode before A. This is because we rely on light coming from the stars to tell us about these events, and light does not travel instantaneously. It travels fast through empty space, at about 300,000km/s, but we cannot tell definitively which star exploded first. What two observers see depends on their relative positions and on their relative speeds.

# In
# 100
### words

In everyday thinking, we picture the space around us as three-dimensional (up-down, left-right, forwards-backwards). Time appears to be something completely different.

Einstein's theory of relativity says we should think of these as merged into a four-dimensional **space-time**. Any "journey" we take is through space-time.

If two people carrying clocks, A and B, set off simultaneously from point X and arrive simultaneously at point Y, we might assume that their two clocks would show that equal times have elapsed. But no. If A has travelled faster than B, their clock will show that, for them, a shorter time has elapsed.

# 17

# Relativistic motion

**WHY IT MATTERS**
More accurately
predicts how objects
move at speeds
approaching the speed
of light

**KEY THINKERS**
Hendrik Lorentz
(1853–1928)
Henri Poincaré
(1854–1912)
Albert Einstein
(1879–1955)

**WHAT COMES NEXT**
Theories incorporating
gravity into relativity

**SEE ALSO**
General relativity, p.78
Hubble's law, p.160

One day people may try to travel to planets orbiting distant stars. This could take thousands of years, unless they can travel in a spacecraft at a speed approaching the speed of light. One of the findings of Einstein's special theory of relativity is that "moving clocks run slow". This effect is known as **time-dilation**, and the effect is greater the closer an object's speed gets to the speed of light. For those travelling in the spacecraft the effect would not be noticeable because it's not just moving clocks that run slow; every process in the human body would also run more slowly. To the space travellers themselves everything would appear normal.

Another consequence of the theory is that, as an object's speed gets closer to the speed of light, its **mass** increases. This will make such a journey even harder to achieve as a greater mass requires a greater force to **accelerate** it.

In
**100**
words

Relativistic effects only become significant when an object's speed approaches the speed of light in free space (a vacuum). Time passes more slowly for a fast-moving object. The **mass** of an object increases as its speed approaches the speed of light.

These effects are too small for us to observe in everyday life, but **time-dilation** has been verified by flying atomic clocks around the world. They are much more significant for subatomic particles moving rapidly in particle accelerators. They are also significant for astronomers observing distant stars that may be moving away from us rapidly as the Universe expands.

# Waves

Physicists describe many phenomena in terms of waves – we're all familiar with sound waves, light waves and electromagnetic waves. Waves are what we see in water, but what have they got to do with sound or light or the electromagnetic spectrum?

Waves are an example of a scientific model: physicists study waves in water and apply what they discover to explain other phenomena. Waves are very different from the particles of the kinetic model of matter – one set of waves can pass straight through another whereas if two particles collide they bounce off one other.

A water wave is a disturbance that travels across the surface of water. The surface of the water moves up and down while the wave travels horizontally – it's a transverse wave. It transfers energy from place to place.

Water waves show the typical characteristics of waves: they reflect off hard surfaces; they **diffract** (spread out) when they pass through a gap or around the edges of a barrier; and they transfer energy.

If two sets of water waves meet, they pass straight through each other unaffected, as seen when the wake of a ship passes through waves on the sea.

**WHY IT MATTERS**
Water waves can act as a simple model for thinking about other wave-like phenomena

**KEY THINKERS**
Christiaan Huygens (1629–1695)
Joseph Fourier (1768–1830)

**WHAT COMES NEXT**
Understanding water waves can help to interpret other wave-like phenomena such as sound and light

**SEE ALSO**
Sound waves, p.38
Frequency and wavelength, p.39
Radio waves, p.40

# Water waves

If you sit in a small boat out at sea you will probably notice that the boat goes gently up and down as waves pass beneath it. That's the kind of wave physicists are thinking of when they picture water waves – not waves breaking on the beach when they reach land.

Water waves can travel great distances without losing their energy. Storms in the South Pacific can create waves that arrive several days later in Hawaii, thousands of kilometres to the north. Surfers can sign up for automated alerts, telling them well in advance when and where to expect big waves.

Water waves are caused when the surface of the water is disturbed. As one patch of water rises up, it pulls up the next one, which pulls up the next, and so on. That's how the wave propagates across the surface.

# 19 Sound waves

If you put your ear to a table and tap one of the legs, the sound reaches your ear. But what exactly is it that travels from the source of the sound to your eardrum?

A solid material is made of particles packed closely together (see page 10). Your finger pushes on one particle, which pushes on the next, and so on. This disturbance travels through the material of the table until it reaches your ear. It isn't matter that has travelled to your ear, it is the disturbance.

Sound can travel through a gas such as air because in a gas, as with solids and liquids, one particle can push another and so on. A vacuum has no particles, so sound cannot pass through.

**WHY IT MATTERS**
Sound waves carry information about their source to our ears

**KEY THINKERS**
Aristotle (384–322BCE)
Joseph Fourier
(1768–1830)
Christian Doppler
(1803–1853)

**WHAT COMES NEXT**
Sound waves have uses in sonar and seismology

**SEE ALSO**
Water waves, p.37
Frequency and wavelength, p.39
Radio waves, p.40

Sound waves are regular mechanical disturbances passing through a material. They carry energy outwards from a vibrating source. The energy is transferred from one particle to the next as the wave travels along.
We can call these disturbances "waves" because they show wave properties: they reflect (think of echoes); they **diffract** or spread out as they pass round barriers (we can hear round corners); and they carry energy.
Two or more sound waves can pass through each other, allowing us to hear the sounds of all the instruments in a band or orchestra at the same time.

**In 100 words**

The frequency of a wave is the number of waves passing a point each second, measured in hertz (Hz). 1 Hz is 1 wave per second. Wavelength is the distance from one wave peak to the next, measured in metres (m). A wave's speed is calculated by multiplying its frequency and wavelength. Sound waves that we can hear have frequencies of 20Hz–20kHz and travel at about 340m/s in air. Visible light waves (see page 42) have much higher frequencies – hundreds of trillions of hertz – and travel at 300 000 000m/s, almost one million times as fast.

**WHY IT MATTERS**
Frequency and wavelength are two important properties of any wave

**KEY THINKERS**
Christiaan Huygens (1629–1695)
James Clerk Maxwell (1831–1879)

**WHAT COMES NEXT**
The frequency of a wave is determined by how it is produced

**SEE ALSO**
Light, p.42

# Frequency and wavelength

Our ears can detect sounds of different pitch. A sound of high pitch is caused by a rapidly vibrating source. More waves reach your ear each second – it has a high frequency.

We can tell one musical instrument from another because the notes they produce are not pure, with a single frequency; instead, they have mixtures of different frequencies – similar to the voices of different people.

A sound with a higher frequency has waves of shorter wavelength – the waves are squashed more closely together because more are being produced each second.

# 21

# Radio waves

In 1901, Italian electrical engineer Guglielmo Marconi showed that he could transmit a signal using radio waves across the Atlantic, from Ireland to Newfoundland. His invention meant that ships crossing the ocean could send out distress calls if they got into trouble. Since then, radio waves have found many uses: for broadcasting, mobile phones, Wi-Fi, Bluetooth and more.

Marconi's radio waves had a frequency of about 1MHz (1 megahertz, or 1 million hertz). Radio waves have a great range of frequencies, up to about 300GHz (gigahertz, or billions of hertz). This range is divided by international convention, so that different frequency ranges are used for different applications. Bluetooth, for example, uses frequencies in the range 2.40–2.48GHz, 5G mobile phones operate in the range 24–54GHz.

The electric and magnetic fields generated in distant astronomical objects such as pulsars fill space with radio waves that can pass through the atmosphere and are detected by the "dish" aerials of radio telescopes.

To produce radio waves, electronic engineers make a varying electric current flow up and down in an antenna (a metal rod or wire), resulting in varying electric and magnetic fields – an **electromagnetic wave** (see page 50) spreading outwards from the transmitting antenna. A second antenna, the receiver, works like the transmitter in reverse. The radio waves produce a varying current in the receiving antenna and this is amplified using electronic circuitry.

The frequency of the current in the receiver is the same as that of the transmitter, and so the receiver captures the information sent by the transmitter.

# 22

# Light

**WHY IT MATTERS**
Light brings us information about our environments

**KEY THINKERS**
Thomas Young
(1773–1829)
Christiaan Huygens
(1629–1695)
James Clerk Maxwell
(1831–1879)

**WHAT COMES NEXT**
Thinking of light as waves enables us to interpret many phenomena

Roughly 80% of the information our senses gather about our surroundings reaches us through our eyes. But what is light? This question preoccupied ancient cultures, and they came up with many different answers.

Perhaps light was produced by our eyes, shining out to see the world around us? Perhaps it was one of the four or five fundamental **elements** from which everything else was made? Perhaps it was waves, travelling through matter, like sound waves. Or perhaps it was streams of particles ("corpuscles") carrying energy from hot objects.

Physicists in the 19th and 20th centuries built on the work of predecessors including Isaac Newton, Thomas Young and Christiaan Huygens to explain many phenomena associated with light. The surprising thing is that sometimes their explanations require us to think of light as **electromagnetic waves** and sometimes as particles (see page 133).

Hot objects produce light. We rely on
light from the Sun (atmospheric
temperature 5,700°C).
Traditional light bulbs have a hot filament that glows,
while light can also be produced by chemical reactions and
by electronic circuits, such as those in light-emitting
diodes (LEDs).
In most cases light is produced when **electrons**, either in
atoms or moving as an electric current, lose energy.
The result is an **electromagnetic wave** spreading
out. When it strikes a surface the wave may
be reflected, transmitted or absorbed.
We see when light is absorbed
by rod and cone cells
in the retinas
of our eyes.

# 23

# Interference

**WHY IT MATTERS**
Interference is a
characteristic
phenomenon of waves

**KEY THINKERS**
Isaac Newton
(1643–1727)
Thomas Young
(1773–1829)

**WHAT COMES NEXT**
A wave's wavelength
can be determined
using interference

**SEE ALSO**
Frequency and
wavelength, p.39

In 1803 English physicist Thomas Young shone a beam of light through two narrow parallel slits onto a screen. He may have expected to see a pair of bright lines on the screen where the light had passed through the slits, but what appeared was a series of equally spaced light and dark stripes called **interference** fringes.

Young's explanation was that light passing through the slits spreads out on the other side. A bright fringe is a point where light waves arrive from the two slits in step with each other so that their effects add together. At a dark fringe the two light waves arrive out of step and cancel out. It's startling to find that two waves of light arriving at the same point can interfere like this to give darkness.

## In 100 words

**Interference** effects are observed when two waves with the same wavelength meet. If the waves arrive in step ("in phase"), their effects add together (constructive interference). If they arrive out of phase, their effects cancel (destructive interference).

Thomas Young used his double-slit experiment to determine the wavelength of light. This convinced many physicists that light is a form of wave.

If two sets of waves meet, their combined effect can be found simply by adding the effects of the individual waves; this is known as superposition. Two waves passing through each other will continue unaffected into the space beyond.

Two identical waves that are exactly in step show constructive interference (top); out of step they show destructive interference (bottom)

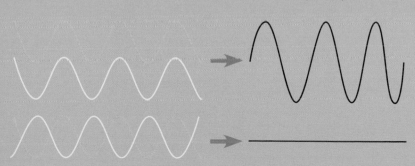

# 24

# Refraction

**WHY IT MATTERS**
Understanding how
light refracts enables
scientists to develop
many technologies

**KEY THINKERS**
Isaac Newton
(1643–1727)
Christiaan Huygens
(1629–1695)

**WHAT COMES NEXT**
Telescopes and
microscopes use many
types of wave other
than visible light

**SEE ALSO**
Light, p.42

A lens is an optical device used to gather and focus rays of light – the word comes from the fact that a lens is shaped like a lentil. Lenses make use of the **refraction** or bending of light, and were probably first used as burning glasses to start fires.

Galileo Galilei's use of lenses in his telescopes caused a revolution in astronomy, enabling us to see the moons of Jupiter and craters on the Moon, breaking the consensus idea that everything in the heavens was "perfect". Similarly, Antonie van Leeuwenhoek and Robert Hooke used microscopes to revolutionise how biologists saw the tiny world of microbes and cells.

The lenses in our eyes are imperfect and deteriorate with age. For those in danger of losing their sight because of the formation of cataracts (when lenses in the eye develop cloudy patches due to fibres and proteins breaking down), the implant of an artificial lens has become something of an everyday miracle.

**Refraction** is the bending of light when it passes from one transparent material to another. Light slows as it passes from, say, air to glass – its speed is reduced by about a third.

A convex lens is thicker in the middle than at the edges. Light waves passing straight through the middle of the lens have less distance to travel than waves passing through the thin edges, but they spend more time inside the glass travelling slowly. Consequently all waves spreading out from a source take the same time to reach their focus on the other side of the lens.

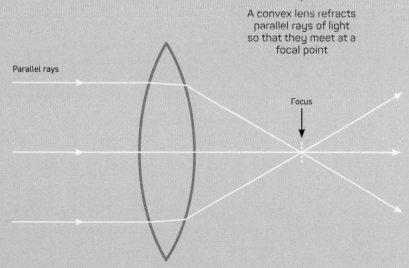

A convex lens refracts parallel rays of light so that they meet at a focal point

Parallel rays

Focus

# 25

# Spectra

**WHY IT MATTERS**
Dispersion allows us to
break down light into
its constituent colours

**KEY THINKERS**
René Descartes
(1596–1650)
Isaac Newton
(1643–1727)
Christiaan Huygens
(1629–1695)
George Robert
Carruthers
(1939–2020)

**WHAT COMES NEXT**
The discoveries of
infrared and ultraviolet
extended the spectrum
of visible light

**SEE ALSO**
The life of a star, p.157

White light passing through a prism produces a spectrum of colours. Traditionally some scientists argued that the prism added colours to white light, but in 1666 Isaac Newton showed that the prism was separating white light into its constituent colours.

The red light at one end of the spectrum is made up of light waves with the lowest frequencies (longest wavelengths); violet light at the other end has the highest frequencies (shortest wavelengths).

The idea of something with a range of characteristics being divided across a spectrum has spread to many other subjects. Pharmacologists develop "broad-spectrum antibiotics" that fight many different bacteria, psychiatrists talk about the "autism spectrum" and social scientists refer to the "political spectrum" from left to right.

Violet
Indigo
Blue
Green
Yellow
Orange
Red

White

When white light enters a prism, the higher frequencies (e.g. violet) are slowed down more than the lower frequencies (e.g. red). This causes them to be refracted through a greater angle. This separation by frequency is called **dispersion** and results in the visible spectrum.

It is often more convenient to produce a spectrum using a diffraction grating – a series of equally spaced parallel slits through which the beam of light **diffracts.** Longer wavelengths are diffracted through greater angles than shorter ones, so a spectrum results.

Many astronomical telescopes incorporate a spectrometer that produces spectra of the light of distant stars.

# 26 The electromagnetic spectrum

The eyes of primates such as humans have evolved to allow us to see light waves with frequencies in the range 420–750THz (terahertz, trillions of hertz). That's useful if you want to see, for example, colourful fruit in a dense green forest.

Other creatures can see outside of this range. Some snakes can see lower frequencies, which means that they can pounce on warm prey even when it's dark. Many insects can see higher frequencies, allowing them to see otherwise invisible markings on flowers.

Nature has evolved to use infrared and ultraviolet radiation beyond the limits of the visible spectrum. These two types of **electromagnetic wave** were identified at the turn of the 19th century. The theoretical work of Scottish physicist James Clerk Maxwell showed that electromagnetic waves of all frequencies travel through empty space at the same speed (300,000,000m/s) – the speed of light. His work unified a great range of phenomena, including light, infrared and ultraviolet, radio waves, X-rays and gamma rays.

**In**
## 100
**words**

**Electromagnetic waves** are oscillating waves of varying electric and magnetic fields. Different types of electromagnetic wave are produced in different ways and have their own properties, but all share this underlying mechanism.

The spectrum of electromagnetic waves stretches from radio waves (low frequencies) through infrared, visible light and ultraviolet to X-rays and gamma rays (high frequencies). The lowest frequencies have the longest wavelengths. All electromagnetic waves can travel through a vacuum – they have no need of a material medium. They travel at the speed of light through empty space. Like all waves, electromagnetic waves transfer energy from place to place.

# 27

# X-rays and gamma rays

At the end of the 19th century, scientists were discovering different types of invisible "rays", spreading out from radioactive rocks, vacuum tubes and so on. Some were particles, others waves. German physicist Wilhelm Röntgen, the discoverer of X-rays, didn't know what his rays were so he called them X-rays, with the X meaning "unknown". Later it became clear that both X-rays and gamma rays are types of **electromagnetic wave**.

X-rays are used in medical diagnosis. Alice Stewart was a British doctor who showed that X-raying expectant mothers can harm the unborn child. The risk is low, but it is worth avoiding, so X-rays are no longer used routinely to examine pregnant women.

> **"**It seemed at first a new kind of invisible light. It was clearly something new, something unrecorded. There is much to do...**"**
>
> **Wilhelm Röntgen**, discoverer of X-rays

X-rays are produced in an X-ray machine where a beam of fast-moving **electrons** crashes into a hard metal target. Their energy is released as high-frequency **electromagnetic waves** — faster electrons give "harder", more penetrating X-rays. X-rays also come from space, where they are produced in violent collisions between astronomical objects.

Gamma rays (γ-rays) are electromagnetic waves produced by radioactive substances as they decay (see page 114). Like X-rays, they are also produced in high-energy astronomical events. By convention, gamma rays are considered to have higher frequencies than X-rays and wavelengths less than a million-millionth of a metre.

# 28

# Scanners

In 1895, German physicist Wilhelm Röntgen made an X-ray image of his wife Anna's hand. This was the first ever medical X-ray. Anna's bones and wedding ring showed up clearly because bone and metal absorb X-rays more than flesh.

A computed tomography (CT) scan can give a more complex image by sending in X-ray beams from different directions. Archaeologists use this technique to examine mummified Egyptian bodies without damaging the delicate wrappings. A computer is required to unravel the large amounts of data this generates to give images of slices through the body. "Tomography" means "seeing slices".

In a magnetic resonance imaging (MRI) scanner the patient is placed in a strong magnetic field and exposed to radio waves. This can provide better images of soft tissues.

**Ultrasound** scans are also good at showing soft tissue. Ultrasound is used because such high-frequency sound waves (beyond our limit of hearing) have short wavelengths, allowing finer detail to be seen. Alongside its medical uses, ultrasound has uses in engineering – for example, to check pipework for corrosion.

MRI and ultrasound have a major advantage over X-rays – the waves used do not damage living tissue.

**In**
**100**
**words**

Scanners use different types of waves to reveal the internal structures of objects. They have uses in medicine, industry, archaeology and other fields. X-rays penetrate some materials more easily than others – for example, bone is more absorbent than flesh. In an MRI scanner, radio waves are absorbed by hydrogen atoms in the body and then re-emitted; this reveals subtle differences in soft tissues. **Ultrasound** waves reflect off the boundaries between tissues; the greater the difference, the stronger the reflection. All these techniques require large computing power to unravel data collected at high speed and produce images of the subject's interior.

# 29

# X-ray crystallography

## WHY IT MATTERS
This technique reveals the structures of materials

## KEY THINKERS
Max von Laue
(1879–1960)
Lawrence Bragg
(1890–1971)
Dorothy Hodgkin
(1910–1994)
Rosalind Franklin
(1920–1958)

## WHAT COMES NEXT
Beams of **electrons** and neutrons can be used in similar ways to X-rays

## SEE ALSO
Interference, p.44
X-rays and gamma rays, p.52

In 1951, English physicist Rosalind Franklin and her New Zealand-born colleague Maurice Wilkins used beams of X-rays to study fibres made of DNA. The X-rays were diffracted by the DNA molecules to produce **interference** patterns which showed that DNA has a helical structure. This result was used by English biophysicist Francis Crick and his American colleague James Watson to make their double-helix **model** of the structure of DNA.

Later, another English scientist, chemist Dorothy Hodgkin used a similar technique to reveal the structures of complex biological molecules including penicillin, insulin and vitamin B12.

"As a scientist Miss Rosalind Franklin was distinguished by extreme clarity and perfection in everything she undertook. Her photographs [of DNA] are among the most beautiful X-ray photographs of any substance ever taken."

**JD Bernal**, crystallographer

Many materials are **crystalline** – their atoms or molecules are arranged in a repeating pattern. X-rays have a wavelength similar to the spacing of atoms in a crystal and for this reason are a suitable probe for investigating the structure of materials at the atomic scale.

A beam of X-rays is **diffracted** when it strikes a crystal so that an **interference** pattern is produced (see page 44). This pattern can be interpreted to give the arrangements of the atoms and their spacings. Scientists have also used X-ray crystallography to reveal the structures of large biomolecules such as insulin and DNA.

# 30

# The Doppler effect

An emergency vehicle, its siren blaring, approaches at speed and then disappears into the distance. As it passes you may notice the pitch of its siren drops – an example of the **Doppler effect**.

The siren emits sound waves whose speed is about 330m/s in air. This means that a stationary source produces a 330m train of waves each second, with the first wave 330m ahead of the last. Now, if the source is moving towards you at 30m/s, the last wave will be only 300m behind the first as it passes you.

So one second's worth of waves is compressed into a shorter distance; the wavelength of the sound is decreased and you hear a higher pitch (a higher frequency). If the source is moving away at 30m/s, the waves will be stretched out over 360m and you will hear a lower pitch.

Scientists observe the same effect with **electromagnetic waves**, but it's harder to observe because the speed is a million times that of sound. Speed cameras direct radio waves at moving vehicles and detect the reflected waves.

The light from a star moving away from the observer is stretched so that its wavelength is longer – this is the origin of the redshift

Observer on Earth

If a source of sound and an observer are moving relative to one another, the observer will detect a change in the frequency of sound waves. The greater the relative speed, the greater the change in frequency (the Doppler shift). The frequency rises if the source and observer are approaching one another and falls if they are moving apart.

The same effect is seen with **electromagnetic waves**. If a star is moving away from an Earth-bound observer, its light spectrum is shifted towards the red end of the spectrum. The speed of the star can be calculated from this **redshift**.

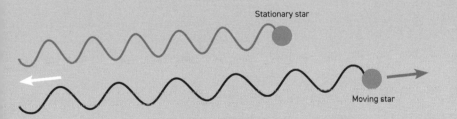

Stationary star

Moving star

# Force Fields

The idea of a field is very widespread in physics – we think of magnetic, electric and gravitational fields. A force field fills three-dimensional space and tells us about the force that will act on an object placed anywhere in the field.

If one magnet is brought close to another they may attract or repel one another, even though they are not touching. This "action at a distance" can seem surprising because we are used to objects exerting forces on each other when they are in contact. We explain it by saying that each magnet is surrounded by a magnetic field.

When astronomers first realised that an orbiting object such as the Moon required a force to hold it in its orbit – another example of action at a distance – they first thought of magnetism as the possible explanation of this force. Now we say that the Moon is held in its orbit by the Earth's gravitational field. While the idea of a force field is rather abstract, many technologies have resulted from physicists' and engineers' ability to think in terms of fields, how to create them and how to control them.

## 31

**In**
**100**
**words**

A magnetic field exists wherever a magnetic pole feels a force. A field can be represented by field lines, showing the direction of the force on a north magnetic pole. Closely spaced lines indicate a stronger region of the field. Magnetic fields can be created in two ways, by a permanent magnet or by an electric current every electric current is surrounded by a magnetic field.

All materials respond to magnetic fields, depending on their atomic structure (see page 105). Ferromagnetic materials such as iron, nickel and cobalt respond most strongly and are used to make permanent magnets

**WHY IT MATTERS**
Magnetic fields are a way of describing the magnetic interaction between magnetic poles, electric currents and magnetic materials

**KEY THINKERS**
William Gilbert (1544–1603)
Hans Christian Ørsted (1777–1851)
James Clerk Maxwell (1831–1879)
Donna Elbert (1928–2019)

**WHAT COMES NEXT**
Two or more magnetic fields can interact to produce motion, the basis of the electric motor

**SEE ALSO**
Atomic structure, p.105

# Magnetic fields

A compass needle is a small permanent magnet. It has two magnetic poles, one of which points roughly to the Earth's North Pole, so we call it the north magnetic pole, while the other is the south magnetic pole.

These magnetic poles attract other magnetic poles of the opposite type, so that north attracts south, and repel poles of the same type. The idea of a magnetic field is used to explain how a magnet can appear to reach out into the space around it.

The Earth itself is like a giant magnet – that's how a compass needle works, as the forces on its poles make it align with the Earth's magnetic field. The Earth's core is metallic, with giant electric currents circulating in it. These currents generate the magnetic field, which extends far out into space.

# 32

# Electric charge

Popular shows of the 18th and 19th centuries demonstrated the effects of static electricity – fairground operators and others would use electrostatic machines to cause sparks to fly through the air, make glass spheres attract one another or ignite fires.

These effects are familiar on a much smaller scale. Run a plastic comb through your hair and you may find your hair stands on end. You may see or hear tiny sparks if you pull off a garment made of a synthetic fabric.

When two different materials are rubbed together, the friction between them may cause them to become electrically charged. The reason is that **electrons**, which have a negative electric charge, are transferred from one material to the other. The material that gains electrons has a negative charge while the other is left with a positive charge.

These oppositely charged materials will attract one another, which is why the comb will attract strands of hair.

## In
# 100
### words

Electric charge is a fundamental property of matter, arising from the particles of which it is made. Subatomic particles may have positive charge (e.g. protons) or negative charge (e.g. **electrons**) or be neutral (e.g. neutrons).

If two uncharged materials are rubbed together, friction transfers electrons from one to the other. Each becomes charged, and the quantities of charge are equal and opposite because the total charge remains constant – it is a **conserved quantity**.

Just as opposite magnetic poles attract, so do opposite electric charges (positive and negative). Two charges with the same sign (both positive or negative) repel each other.

# 33

# Electric fields

Lightning is a dramatic effect of static electricity. When the lower part of a thundercloud gains an electric charge it is discharged to the ground in the form of a spark – a large electric current.

Inside a thundercloud, ice crystals are moved around by **convection** currents. They rub against each other so that **electrons** are transferred from smaller crystals to larger ones, which sink to the bottom of the cloud. In this way the bottom of the cloud becomes negatively charged while the top is positive.

There is now a strong electric field between the cloud and the ground and this produces the spark, the electric current in the air. Negative charges move downwards, positive charges upwards. If you are beneath the cloud, your hair may stand on end as it responds to the electric field.

Uncharged matter is described as neutral. It has
equal quantities of positive and negative electric
charge, which are attracted to each other and
so stick together.

If these charges are pulled apart – for example by the
frictional force caused when two different materials are
rubbed together – an electric field is created. A force
acts on any charged object placed in the field, and the
stronger the force, the stronger the field.

Electric and magnetic fields together can be
thought of as a single field, the electromagnetic
field. **Electromagnetic waves** are regular
variations in this field (see page 50).

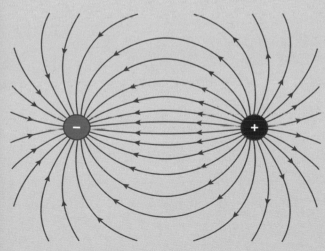

An electric field,
created by two
opposing forces
acting upon one
another

65

# 34

# Electric current

In a famous experiment in 1752, American scientist Benjamin Franklin showed that lightning is a form of electricity by flying a kite in a thunderstorm. This is dangerous – at least one experimenter has died trying to reproduce Franklin's work.

A lightning flash is a large and uncontrolled electric current. **Electrons** are being pushed through the air by the electric field between the cloud and the ground below. Franklin went on to invent the lightning conductor, a metal strip down the side of a building which directs the lightning current into the ground. His invention has saved many lives.

Metals are different from other materials because many of the electrons that are part of the atoms of which the metal is made can move around freely inside it, rather than being bound tightly to their atoms. These **"free electrons"** are what make metals good **conductors** of electricity. The voltage indicated on a battery or other source of electrical power is a measure of the electrical field it creates when connected in a circuit. This field pushes the free electrons through the metal, and this movement of electrical charge is what we call an electric current.

An electric current is the movement of electric charge. In an electric circuit, a battery or power supply provides the voltage needed to create an electric current: when the two ends (positive and negative) of the supply are connected by a metal wire, an electric field is created within the wire. This pushes **electrons** away from the negative end towards the positive ond   thorc io on clectric current in the wire.

A supply with a bigger voltage creates a stronger push, so there is a bigger current. More electrons (so more charge) pass each point in the circuit each second.

# 35

# Electro-magnetism

The connection between electricity and magnetism was discovered in 1820 by Danish scientist Hans Christian Ørsted. He was giving a public lecture on electricity when he noticed that, whenever he connected a circuit to make a current flow in a wire, a nearby compass needle moved round. He realised that this must mean that the current was somehow creating a magnetic field. We now know that whenever there is an electric current there is a magnetic field around it.

This is the basis of an electromagnet, a length of wire formed into a coil to concentrate the magnetic field when there is a current in the wire. To make an electric motor, you need two electromagnets (or one and a permanent magnet). Then the two magnetic fields attract or repel each other, causing one of the coils to rotate.

**WHY IT MATTERS**
An electric current is surrounded by a magnetic field

**KEY THINKERS**
Hans Christian Ørsted (1777–1851)
Michael Faraday (1791–1867)
James Clerk Maxwell (1831–1879)

**WHAT COMES NEXT**
The magnetic effect of a current is used in electromagnetic devices such as motors and MRI scanners

**SEE ALSO**
Radio waves, p.40

## In 100 words

Electricity and magnetism are intimately connected. An electric current is created by an electric field, which causes electric charges to move. Every electric current is surrounded by a magnetic field. A bigger current gives a stronger magnetic field. An electromagnet is a coil of wire with a current in it, resulting in opposite magnetic poles at either end. An electric motor turns when the two magnetic fields interact with each other.
A varying current produces a varying magnetic field which spreads out from the wire. This is the basis of the production of radio waves (see page 40).

## 36

## In **100** words

Electromagnetic induction is production of an electric current in a **conductor** such as a metal wire. It occurs when the conductor is placed in a changing magnetic field, or when the conductor is moved across a steady magnetic field. In a generator such as a dynamo, the changing magnetic field is produced by a rotating electromagnet. As the electromagnet turns, the induced current in the conductor flows, first one way and then the other. This is an alternating current. Connecting the generator into a circuit allows the current to transfer energy from the generator to other components in the circuit.

**WHY IT MATTERS**
Large-scale electricity supply systems mostly make use of electromagnetic induction

**KEY THINKERS**
Michael Faraday (1791–1867)
James Clerk Maxwell (1831–1879)
Emil Lenz (1804–1865)

**WHAT COMES NEXT**
Photovoltaics (solar cells) provide an alternative means of generating electricity

**SEE ALSO**
Electromagnetism, p.68

# Electro-magnetic induction

An electric vehicle uses electric current supplied by a battery to make a motor turn, moving the car forwards. When the driver brakes, the system works in reverse, generating a current that recharges the battery. This represents a great saving in energy; older vehicles use friction to slow down, producing heat that is wasted as it is dispersed in the environment.

The giant dynamos in power stations and wind turbines work like electric motors in reverse. In a motor, a current provides two magnetic fields whose interaction produces motion. In a generator, a rotating magnetic field produces a current in an electrical **conductor**.

# 37

# Particle accelerators

Physicists have found many uses for beams of fast-moving charged particles – **electrons**, protons or **ions**. For example, proton beam therapy is used in the treatment of some cancers and traditional television sets used a beam of electrons in a vacuum tube known as a cathode ray tube.

The largest particle accelerator in the world is the Large Hadron Collider (LHC) near Geneva, Switzerland, where the **Higgs boson** (an **elementary particle** that gives other particles their **mass**) was detected in 2012. Beams of protons or lead ions travel around the 27km-circumference circular underground tunnel of the LHC: two beams of protons, travelling in opposite directions at just 3m/s less than the speed of light, collide. Electronic instruments examine the resulting debris in the hope of finding out more about the fundamental nature of matter.

Before the LHC started to operate, scare stories appeared suggesting that it might create a black hole that would destroy the world. If you are reading this book, that hasn't happened yet.

In a particle accelerator, strong electric fields are used to **accelerate** charged particles. The energy of an accelerated particle is often given in electron-volts (eV); an **electron** with an energy of 1 MeV (mega-electron-volt) has been accelerated through a potential difference of 1 million volts.

Magnetic fields are used to make a beam of particles follow a curved path. The faster the particles are moving and the greater their **mass**, the stronger the magnetic field must be.

Some accelerators produce beams of high-energy X-rays. Deflecting a beam of electrons using a magnetic field causes them to lose energy as X-rays.

" I do theoretical particle physics. We're trying to understand the most basic structure of matter. And the way you do that is you have to look at really small distances. And to get to small distances, you need high energies. "

**Lisa Randall**, theoretical physicist

# 38

# Gravitation

**WHY IT MATTERS**
Gravitation is one of
the fundamental forces
of nature

**KEY THINKERS**
Galileo Galilei
(1564–1642)
Isaac Newton
(1643–1727)
Albert Einstein
(1879–1955)

**WHAT COMES NEXT**
Gravitation has had a
major influence on
the evolution of the
Universe

Throughout our lives we experience the Earth's gravitational pull. It's what makes toddlers topple over when they are learning to walk. It's what makes pole vaulters drop to the ground after they have cleared the bar. It's what makes rain fall from the clouds. All of these things – toddlers, pole vaulters and raindrops – have **mass** (see page 12) and it's the Earth's gravitational pull on an object's mass that causes it to fall.

The Earth is a massive object – its mass is about 6 trillion trillion kilograms. This mass pulls on every object with mass on its surface or anywhere near it. The attractive **gravitational force** of the Earth on an object is what we call the object's weight.

"I have explained the phenomena of the heavens and of our sea by the force of gravity, but I have not yet assigned a cause to gravity."

**Isaac Newton**, physicist

Gravitation is one of the four fundamental forces that act on matter. It is a force arising from the masses of two objects and causes them to be attracted to one another. Newton's theory of universal gravitation explained how the planets were held in their orbits by the Sun's gravitation, an example of action-at-a-distance. Because of the Earth's great **mass**, it exerts an appreciable pull on an object on or near its surface, a force which we call weight. This force is directed towards the centre of the Earth. The object exerts an equal and opposite pull on the Earth.

# 39

# Gravitational fields

**WHY IT MATTERS**
Gravitational fields explain the action at a distance between astronomical objects

**KEY THINKERS**
Isaac Newton (1643–1727)
Albert Einstein (1879–1955)

**WHAT COMES NEXT**
Einstein's General Theory of Relativity recast the way physicists think of gravity

**SEE ALSO**
Gravitation, p.72
General relativity, p.78

Astronauts who have visited the Moon have had the unusual experience of walking in a **gravitational field** that is much weaker than the Earth's. The Moon's **mass** is about one-eightieth of the Earth's and its surface gravity is about one-sixth of the Earth's. This accounts for the lack of atmosphere on the Moon. Any atmosphere it may have once had has since escaped, because molecules moving at typical speeds of 400m/s gradually leave the Moon's surface and disappear into space.

The Earth's gravitational field extends far into space; it's what keeps the Moon in its orbit. In the same way, the Sun's gravitational field keeps the Earth in its orbit. The Sun and Earth pull on each other with equal and opposite forces, so that both orbit around a point on the line joining their centres, much closer to the Sun than the Earth because its mass is so much greater.

If you climb a high mountain, you will find that gravity is weaker, because you are further from the centre of the Earth. If you could descend a long way into the Earth, you would find that its gravity was weaker, because only the mass of the Earth that is closer to its centre contributes to its pull on you.

**In 100 words**

There is a **gravitational field** around any object
with **mass**. This is only significant for objects
with large mass, such as astronomical objects.
There is a gravitational attraction between two
human-sized objects, but it is negligible.
The gravitational field of a star or planet extends far
out into space; in principle, it extends to infinity, but
it gets weaker with distance. Newton showed that
the motion of the planets in the solar system
could be explained by an inverse square law:
doubling the distance of a planet from the
Sun reduces the strength of the
gravitational field to
one quarter.

# 40

# Black holes

**WHY IT MATTERS**
Black holes constitute
a significant fraction of
the mass of the
Universe

**KEY THINKERS**
Albert Einstein
(1879–1955)
Karl Schwarzschild
(1873–1916)
Stephen Hawking
(1942–2018)
Gabriela González
(1965–)

**WHAT COMES NEXT**
Primordial black holes
formed in the early
universe may help to
explain anomalies in its
expansion

**SEE ALSO**
Gravitational
fields, p.74
$E = mc^2$, p.102
The life of a star, p.157
The Big Bang, p.162

At the end of its life, when all of its hydrogen fuel has been used up, a massive star (with several times the **mass** of the Sun) may form a black hole. The star's own **gravitational field** is so strong that it collapses inwards to form an exceedingly dense object, so dense that even light cannot escape from it.

Although we may think of light as "pure energy" with no **mass**, anything with energy does have mass. This is expressed by Einstein's equation, $E = mc^2$ (see page 102). Any light trying to leave a black hole will experience a gravitational pull acting on its mass which slows it down until it falls back inside – it cannot escape.

The first image of a black hole was made by the Event Horizon Telescope (a network of Earth-based radio telescopes) in 2019. Although we cannot see a black hole directly, radio waves from stars behind it will be bent by its gravitational field as they travel towards us and this allows an image of the black hole to be built up.

A black hole is an extremely dense object possibly formed when a massive star collapsed or during the Big Bang (see page 162). It is so called because light emitted within a radius called the **event horizon** cannot escape the black hole's **gravitational field**.

For a distant observer, processes close to a black hole appear to slow down. This is **gravitational time dilation**; one result is that the frequency of light decreases, the **gravitational redshift**.

Most galaxies are thought to have a supermassive black hole at their centre, formed by the merging of multiple black holes.

"Black holes are a gift, both physically and theoretically. They are detectable on the farthest reaches of the observable universe. They anchor galaxies, providing a centre for our own galactic pinwheel and possibly every other island of stars. And theoretically, they provide a laboratory for the exploration of the farthest reaches of the mind."

**Janna Levin**, physicist

# 41

# General relativity

Picture yourself strapped into a rocket accelerating you upwards into space. Your seat pushes up on you, so that you feel as if your weight has greatly increased. Einstein realised that it is impossible to tell whether you are being accelerated by an external force or the force of gravity has increased.

In his general theory, published in 1915, Einstein explained the orbits of the planets by saying that the **mass** of the Sun distorts **space-time** (see page 32) so that it becomes curved; an orbiting planet is simply following this curve. The theory also predicted the existence of gravitational waves, periodic variations in the distortion of space-time, spreading out at the speed of light from violent events in the cosmos such as collisions between black holes. Gravitational waves were first detected in 2017.

**WHY IT MATTERS**
Einstein united ideas about motion and electromagnetism with ideas about gravity

**KEY THINKERS**
Albert Einstein (1879–1955)
Emmy Noether (1882–1935)
Arthur Eddington (1882–1944)
Subrahmanyan Chandrasekhar (1910–1995)
Marie-Antoinette Tonnelat (1912–1980)

**WHAT COMES NEXT**
Einstein's theories of relativity have yet to be made consistent with quantum physics

**SEE ALSO**
Space-time, p.32
Relativistic motion, p.34
Space travel, p.148

## In 100 words

The general theory of relativity replaced Newton's law of universal gravitation. The theory suggests that objects with **mass** distort **space-time**; the greater the mass, the greater the distortion. Other objects with mass move in this distorted space-time. The theory successfully accounted for the details of Mercury's orbit around the Sun and the **gravitational redshift** of light from distant stars. It predicted the existence of black holes and gravitational waves. Einstein's theories are also consistent with Maxwell's equations of electromagnetism. These show that it is the relative motion between **conductors** and electric and magnetic fields that give rise to electromagnetic induction.

# Energy

We can think of a lump of coal as a store of energy: burn it with oxygen and it can keep us warm; burn it in a steam engine and it can move a train on rails. And yet when we look at a piece of coal it seems to be an inert, black mass. It may have been buried underground for 360 million years. Where is the energy that is released by burning?

The word "energy" comes from the Greek word *energeia*, meaning "activity", and yet there is no evident activity in a lump of coal. Rather, coal stores energy that is the result of activity in the past – photosynthesis by trees millions of years ago. Isaac Newton didn't think in terms of energy – the scientific idea had not been developed in his day. Today, rather than thinking of energy as some kind of "stuff" that makes things happen, scientists regard it as more of an accounting tool that enables us to calculate the answers to questions such as "How high will this stone rise when I throw it?", "How hot will this solution become when its ingredients react?" or "Why can't I power my car with a wind turbine?'

# 42

# Energy resources

An adult human body requires about 100 *joules* (J) of energy each second to function normally. (The unit of energy, the joule, is named after English physicist James Prescott Joule, who famously chose to honeymoon in the French Alps so that he could measure temperature differences between the top and bottom of a waterfall.) But if you have a physically demanding job such as working on a farm you will require significantly more. We get this energy from our food.

In the pre-industrial era people burned wood for heating and cooking. They used animals (which also had to be fed) to do much of the hard work of farming.

Since the Industrial Revolution, our energy consumption has vastly increased. The factories that drove the revolution in our way of life were powered by coal-fired steam engines. Today we need energy to heat and light our homes, for travel and construction, for entertainment and leisure. In developed nations, the rate at which the average individual uses energy is between 3,000 and 10,000 J per second – that's up to 100 times what our food supplies.

Much of this energy is supplied by fossil fuels – coal, oil and natural gas. Increasingly we use the energy stored in moving air (wind power) and water (hydro power). All of these stores of energy are derived ultimately from sunlight. We also use sunlight directly, in solar cells.

**In 100 words**

Humans use energy resources for a multitude of purposes, allowing us to do many different things which we could not do simply by using our bodies. Most of these resources are stores of energy that has come from sunlight, either ancient sunlight (fossil fuels), or recent sunlight (solar cells, wind power, hydroelectricity, biomass etc.). Nuclear and geothermal power rely on energy stored in **elements** such as uranium buried in the Earth, but these represent only a small fraction of worldwide consumption.
Technologies such as power stations, nuclear reactors, photovoltaics and turbines are needed to take advantage of these energy resources

**"**It is sunlight in modified form which turns all the windmills and water wheels and the machinery which they drive. It is the energy derived from coal and petroleum (fossil sunlight) which propels our steam and gas engines, our locomotives and automobiles. Food is simply sunlight in cold storage.**"**
**John Harvey Kellogg**, inventor of cornflakes

# 43

# Global warming

Early in the 19th century French scientists such as Joseph Fourier raised the alarm about Britain's use of coal to fuel the Industrial Revolution. They pointed out that coal reserves had been laid down tens or hundreds of millions of years earlier when the Earth was warmer because the atmosphere at the time contained a high level of carbon dioxide. There was a risk that burning a significant fraction of the Earth's coal reserves could return our climate to that of the past. (France had less coal available, making it harder to industrialise, and this may have influenced the thinking of French scientists.)

Two centuries later we know that they were right. The fraction of carbon dioxide in the atmosphere before the Industrial Revolution was 280 parts per million (ppm). By 2020 it had increased by half, to 420 ppm. These fractions may seem small but the presence of carbon dioxide in such quantities is what gives the atmosphere much of its ability to keep the Earth warm. The Moon, which has no atmosphere, is about 30°C (50°F) colder than the Earth, on average.

The atmosphere allows visible light and ultraviolet to reach the Earth's surface, but absorbs infrared radiating out into space. The result is an increase in the Earth's average surface temperature (the greenhouse effect).

John Tyndall measured absorption of heat by different gases and Svante Arrhenius calculated how much this raised the atmospheric temperature. Increasing levels of carbon dioxide arising from the burning of fossil fuels have produced a global warming effect of 1.5°C (2.7°F). This is likely to increase.

Resulting climate change effects include rising temperatures, stormier weather and the melting of polar ice – with consequent increases in sea levels.

# 44

# Kinetic energy

**Kinetic energy** is the energy of a moving object. A moving bus, a ball thrown through the air, a planet orbiting its star – all these have kinetic energy.

Say a snooker player strikes the white ball so that it hits a stationary red ball. As the balls diverge, each is moving more slowly than the white ball was originally; its kinetic energy has been shared between them. A clever player can make the white ball score a direct hit so that it stops dead, and the red moves off at the speed of the white; all the white ball's kinetic energy has been transferred to the red ball. It is impossible for the red to move off faster – kinetic energy would have appeared from nowhere.

We can picture the molecules of a gaseous material such as the atmosphere in a similar way. Each molecule has kinetic energy. As they collide with each other their kinetic energy is shared between them. The warming of the atmosphere has increased the average molecular kinetic energy by about 0.5%.

**Kinetic energy** is the energy of an object due to its motion. It depends on the **mass** and the speed of the object. An object with more mass and moving faster has more kinetic energy.

An object can be given kinetic energy by exerting a force on it. A footballer kicks a ball; while the player's boot is in contact with the ball, it **accelerates**. Its kinetic energy increases, transferred to it by the force of the boot.

Similarly, an object can lose kinetic energy when a force such as friction acts on it. Energy is transferred to the surroundings.

# 45

# Potential energy

A weightlifter raises weights above her head. The weights have been given gravitational **potential energy**, transferred to them by the lifting force of the weightlifter. If she drops the weights they will fall, possibly causing damage as their energy is transferred to the floor.

Weights are lifted against the force of gravity. Whenever an object is moved against an opposing force, it is being given potential energy.

Another example is **photosynthesis**. Plants use the energy of sunlight to pull molecules of water and carbon dioxide apart. The opposing force is the electrical force that holds the original molecules together. Then they put the atoms back together in a different arrangement to make molecules of sugar and oxygen. In these molecules the atoms are less tightly bound together. They have **chemical potential energy**, so they act as energy stores for the plant. The plant can recover this energy later in the process of respiration, in which the atoms return to their original and more tightly bound conformation as molecules of water and carbon dioxide. We, too, have the benefit of this energy when we eat plants and breathe in oxygen.

**Potential energy** is the energy an object has because of its position in relation to other objects. For example, gravitational potential energy is the energy of an object raised in a **gravitational field.** Pulling apart two magnets whose opposite poles are attracting one another increases their magnetic potential energy. Pulling apart two opposite electric charges increases their electrostatic potential energy, as does pushing together two like electric charges.

**Chemical potential energy** is the energy stored in a chemical substance. The energy can be released in a chemical reaction, after which the constituent atoms are more tightly bound together.

# 46

# Doing work

**WHY IT MATTERS**
Forces transfer energy
from one object to
another when they
do work

**WHAT COMES NEXT**
Understanding how
forces transfer energy
is vital in designing
machines

Like many words in science, "work" has a special meaning in physics. Here are two examples of forces "doing work":

A tennis player serves at the start of a match. The force of the racket on the ball pushes the ball for a short distance, causing it to **accelerate**; the ball has been given energy, in this case **kinetic energy**. The greater the force and the greater the distance it moves while in contact with the ball, the greater the ball's kinetic energy.

If you lift a book from the floor onto a shelf, the force you use to lift the book does work. It increases the gravitational **potential energy** of the book. If you use a force of 3N to lift the book vertically through 2m, the force does 3N × 2m = 6J of work and the book has been given 6J of gravitational potential energy.

In **100** words

A force does work when it moves through a distance. The work done (in joules, J) is given by work done = force × distance moved by the force. In doing work, a force transfers energy to the object it moves. The energy transferred is equal to the work done by the force.

No work is done, and no energy transferred, by a force if it doesn't move. For example, the upward push of the floor on someone's feet stops them falling through the floor but, since the force doesn't move, it does no work and no energy is transferred.

# 47 Heat transfer

A man is cooking using a traditional method. He heats a rock on an open fire, then places the hot rock in a container of water along with the food to be cooked; the water is heated by the rock and, in turn, this heats the food – the energy of the hot rock has been shared with the water and the food.

We rely on heat transfer in many ways. Homes in cold climates have radiators from which energy is transferred to the rooms. On a giant scale, we rely on heat transfer of energy from the Sun (surface temperature 5,500°C [9,900°F]).

Heat transfer can also be a problem. Homes may have to be insulated to reduce the loss of energy on a cold day. In hot weather, heat transfer can cause homes to become too hot.

**WHY IT MATTERS**
Heat transfer, like a force doing work, is a mechanism of energy transfer

**KEY THINKERS**
Sadi Carnot (1796–1832)
William Rankine (1820–1872)
William Thomson (1824–1907)

**WHAT COMES NEXT**
The ideas of work and heat transfer led to the science of thermodynamics

**SEE ALSO**
The Laws of Thermodynamics, p.98

## In 100 words

Heat transfer is the movement of energy as a result of a temperature difference. Energy is transferred from a hotter to a colder place. A hot object can be described as a thermal store of energy. It can transfer energy to its surroundings by a variety of mechanisms including **conduction**, **convection** and radiation. The **heat capacity** of an object is the energy needed to raise its temperature by 1°C. Knowing this, we can calculate the energy stored in the object at a given temperature above that of its surroundings. This is how much energy it can lose by heat transfer.

## 48

If two objects are in thermal contact, energy will transfer from the hotter to the colder until they are in thermal equilibrium – when there is no net transfer of energy between them. Temperature difference tells us the direction in which energy is transferred. All objects are constantly losing energy through the emission of infrared radiation, a form of electromagnetic radiation (see page 50). They also absorb infrared radiation from their surroundings. If an object and its surroundings are at the same temperature, there will be no net transfer of energy between them; the object's temperature will remain constant.

**WHY IT MATTERS**
Temperature difference determines the direction of heat transfer

**KEY THINKERS**
James Clerk Maxwell (1831–1879)
Ralph Fowler (1889–1944)
Mária Telkes (1900–1995)

**WHAT COMES NEXT**
Temperature differences make heat engines work

**SEE ALSO**
Heat engines, p.100

# Thermal equilibrium

If you place a cold thermometer in a hot cup of tea, its reading will rapidly rise until it reaches the temperature of the tea. The thermometer and the tea are in thermal contact; energy transfers from the tea to the thermometer until they are in thermal equilibrium. Energy transfers back and forth between them at equal rates.

At night it gets cold not simply because the dark side of the Earth is facing out into the darkness of space where the average temperature is about -270°C (-454°F). The Earth cools as infrared radiation spreads out from its surface, as it does from any object whose temperature is above **absolute zero** (-273°C or 0K). If the Earth turned more slowly, night-time would be longer and colder.

# 49

# The principle of conservation of energy

**WHY IT MATTERS**
The principle tells us
that energy can be
neither created nor
destroyed

**KEY THINKERS**
Émilie du Châtelet
(1706–1749)
Thomas Young
(1773–1829)
James Joule
(1818–1889)
Emmy Noether
(1882–1935)

**WHAT COMES NEXT**
The Laws of
Thermodynamics

**SEE ALSO**
The Laws of
Thermodynamics, p.98

Throw a ball vertically upwards and it will soon return to your hand. When you catch it, it will be travelling at the same speed as when it left your hand, but in the opposite direction.

We can describe this in terms of energy changes. The force of your hand has done work on the ball, giving it **kinetic energy**. As it rises, the Earth's gravitational pull does work on it, slowing it down. At its highest point it is instantaneously stationary. It has no kinetic energy but it does have gravitational **potential energy**. As it falls, gravity continues to do work on it, speeding it up so that its kinetic energy increases as its potential energy decreases.

As the ball rises and falls, its kinetic energy changes to potential energy and back again. At each point in its trajectory, the sum of the ball's kinetic and potential energies is the same. This is an example of the conservation of energy – energy can't disappear or appear from nowhere.

Keeping track of the energy changes that happen in complex systems can be tricky, but provided no energy is escaping we can be sure that the total energy in a system is constant – energy is conserved.

The principle of conservation of energy says that, within a closed system, the total energy is constant. A "closed system" is one from which no energy is lost as the system changes Because we can calculate the **kinetic** and **potential energies** of objects, we can use the principle to predict the outcome of changes: How high will a rocket rise? How hot will a tank of water get? How fast will a spacecraft move?
In a sense this tells us what energy is. It is a calculated quantity whose value remains constant for a closed system.

# 50

# Entropy

Take a box and some marbles. Put a layer of blue marbles in the bottom of the box. Put a layer of red marbles on top. This is an orderly arrangement.

Now shake the box. The two layers mix, producing a random, disorderly arrangement. Keep shaking the box ... You would be very surprised to look inside and find that they had gone back to their original arrangement, or even to find a layer of blue on top of a layer of red.

This tendency to randomness is something we see everywhere in nature. Hot objects cool down; their energy spreads around, becoming less concentrated and so less useful. Increasing randomness can be thought of as an indication of **"time's arrow"**. Time moves in the direction of increasing randomness.

> **"**You may see a cup of tea fall off a table and break into pieces on the floor... But you will never see the cup gather itself back together and jump back on the table. The increase of disorder, or entropy, is what distinguishes the past from the future, giving a direction to time.**"**
>
> **Stephen Hawking**, cosmologist

In
**100**
words

When objects interact, the forces
between them (such as friction) often lead
to energy being dissipated as thermal energy.
This is the random motion of particles and represents
an increase in **entropy**.
Entropy is a measure of the randomness of the arrangement
of particles in a system. An ordered system has low entropy.
As a system becomes more disordered its entropy increases.
The entropy of the universe as a whole is constantly
increasing. An interaction may appear to
produce greater order and hence reduce
entropy. However, when the
surroundings are taken into
account, entropy is always
found to increase.

# 51

# Absolute zero

## WHY IT MATTERS
Absolute zero helps to define a temperature scale that is not dependent on physical properties of materials, such as the freezing point of water

## KEY THINKERS
Sadi Carnot
(1796–1832)
William Thomson
(1824–1907)
Ludwig Boltzmann
(1844–1906)

## WHAT COMES NEXT
Temperatures on the kelvin scale are required when calculating the theoretical efficiency of a heat engine

## SEE ALSO
Charles's law, p.14

Pure water freezes to become ice at 0°C (32°F). The freezer section of a domestic fridge-freezer keeps food at -20°C (-4°F), while commercial freezers used for storing vaccines, donor tissue samples and other biological materials may go down to -80°C (-112°F). That's cold, but it's still almost 200 degrees above the coldest temperature, **absolute zero**.

The molecules of a gas move around freely. At room temperature, the molecules of oxygen and nitrogen in air have an average speed close to 400m/s, slightly greater than the speed of sound in air. Cool the air down and the molecules move more slowly. It's not hard to imagine that, if you keep cooling the air, eventually the molecules will come to a stop.

Ludwig Boltzmann made this relationship mathematical. He suggested that the average **kinetic energy** of the molecules of a gas was proportional to the temperature of the gas. The lowest possible temperature is when the molecules have no kinetic energy – they have stopped moving. This is the temperature we now call absolute zero, equal to -273.15° on the Celsius scale, -459.67° in Fahrenheit and 0 on the Kelvin scale. (In practice, all gases become liquids before their temperature reaches 0K.)

**Absolute zero** is the temperature at which
materials have their lowest values of energy
and **entropy**. Although motion remains, no further
energy can be extracted from the material.
Absolute zero is the zero-point on the Kelvin scale,
written as 0K. The unit K (or kelvin) is equal to the unit
on the Celsius scale.
It is impossible to reach this temperature in practice.
In cooling an object, only a fraction of its energy
can be removed. For each cooling cycle
some of the energy is removed and
the temperature drops, but a
fraction of the energy remains.

# 52

# The Laws of Thermo-dynamics

In 1959 physicist, novelist and politician C. P. Snow complained that, while it was assumed that an educated person would know the plays of Shakespeare, few could explain the significance of the second law of thermodynamics. There are four Laws of Thermodynamics. The most fundamental law was added to the list after the others were well established, and so is known as the "zeroth" law.

The four laws (see box) can be expressed like this. The zeroth law says that if objects A and B are at the same temperature, and B and C are also, then A and C are, too. Then the first law explains that energy is conserved; it can be neither created nor destroyed. The second law says that heat flows from hot objects to cold objects, and tends to be dissipated as heat – the disorder of the system increases. And the third law says that removing energy from a system reduces its temperature, but **absolute zero** can never be reached.

**ZEROTH LAW:** If A and B are in thermal equilibrium with one another, and B and C are also, then A and C are in thermal equilibrium.

**FIRST LAW:** If a force does work (W) on a body and an amount of energy (H) is transferred to it by thermal transfer, its energy will increase by the sum of these, W + H.

**SECOND LAW:** Thermal energy transfer occurs down a temperature gradient from hot to cold. **Entropy** increases.

**THIRD LAW:** At **absolute zero** a system has minimum values of energy and entropy, although this is unattainable in practice.

"The history of thermodynamics is a story of people and concepts. The cast of characters is large. At least ten scientists played major roles in creating thermodynamics, and their work spanned more than a century. The list of concepts, on the other hand, is surprisingly small; there are just three leading concepts in thermodynamics: energy, entropy, and absolute temperature."

**William H. Cropper**, author of *Great Physicists*

# 53

# Heat engines

**WHY IT MATTERS**
Heat engines burn fuel
to produce useful work

**KEY THINKERS**
James Watt
(1736 1819)
Sadi Carnot
(1796–1832)
William Thomson
(1824–1907)

**WHAT COMES NEXT**
Because of global
warming, energy users
are turning away from
heat engines

**SEE ALSO**
The Laws of
Thermodynamics, p.98

The Laws of Thermodynamics were developed at the time of the Industrial Revolution and helped scientists and engineers improve the design of the engines that burned fuels to power machinery that generated great wealth.

They installed steam engines in mines, factories and locomotives. A steam engine is an example of a heat engine – any engine that makes use of the heat from burning fuel. Modern examples include the internal combustion engine used in vehicles, jet engines used by aircraft, and even nuclear power stations.

The Laws of Thermodynamics tell us that it is inevitable that much of the energy released by a hot fuel will be dissipated as waste heat to the surroundings. In other words, the efficiency of a heat engine is considerably less than 100%. This is not true for other types of engine, such as electric generators and motors, which can be much more efficient in their use of the energy supplied to them.

In a heat engine, energy is transferred from a hot reservoir (such as a boiler, where steam is generated) to a cold reservoir (such as a condenser, where steam is cooled back to water). As the energy is transferred, some becomes the mechanical energy of moving pistons and turning wheels.

It is impossible to turn all the random motion of a hot reservoir into useful motion: the efficiency of a heat engine is limited. Typically, a heat engine's efficiency is below 40% – that is, more than 60% of the energy supplied by its fuel is lost as waste heat.

# 54

$$E = mc^2$$

**WHY IT MATTERS**
The mass of an object
changes as its energy
changes

**WHAT COMES NEXT**
Mass changes during
nuclear reactions can
be explained

If you pick up a book from the floor and put it on a shelf, you have increased its gravitational **potential energy**. But, looking at the book, you won't see any difference. This is because the change in **mass** is very, very small.

Einstein's equation $E = mc^2$ tells us that, if you have increased the book's energy by an amount ($E$), you will have also increased its mass by an amount ($m$). To calculate the change in mass we divide the energy change by the square of the speed of light ($c$), which is 300,000,000m/s. Suppose the book's energy increases by 10J. Its increase in mass is $10 / (300,000,000)^2$ – which works out at less than a trillionth of a gram, so small a change that you can't notice it.

A hot object will cool down by emitting infrared radiation. Its energy decreases as it cools and so its mass also decreases by a tiny amount.

Albert Einstein deduced the equation $E = mc^2$ as part of his special theory of relativity (see page 32). It is known as the mass-energy equivalence formula.

Any increase in the energy ($E$) of an object is accompanied by an increase in its **mass** ($m$). As a spacecraft **accelerates**, its **kinetic energy** increases and so does its mass.

Changes in $E$ and $m$ are related by $c^2$, the square of the speed of light in a vacuum. Because $c^2$ is a very large quantity, a small energy change results in a minuscule change in mass, negligible in everyday life.

"You may not feel outstandingly robust, but if you are an average-sized adult you will contain within your modest frame no less than $7 \times 10^{18}$ joules of potential energy—enough to explode with the force of thirty very large hydrogen bombs, assuming you knew how to liberate it and really wished to make a point."

**Bill Bryson**, writer

# Atomic Physics

The 20th century can be thought of as the century of the atom. Once Albert Einstein had explained Brownian motion (see page 9) no scientist could reject the idea that matter was made of atoms. Then the nucleus of the atoms was discovered by Ernest Rutherford and his colleagues. The structure of the atom was explained in terms of subatomic particles – protons, neutrons and electrons.

The forces within the nucleus explain radioactivity, fission and fusion. Nuclear fission and fusion are the mechanisms used in the nuclear weapons that came to dominate international politics after the Second World War (1939–1945).

Gradually scientists discovered more and more particles – and today we have the standard model, which goes a long way to uniting all of these observations.

It is still incomplete and may, like all ideas in physics, be overturned through future discoveries. Watch this space!

Every atom has a central **nucleus** which has most of the **mass** of the atom. It is made of protons and neutrons. These are particles with similar masses; a proton has a positive electrical charge. Orbiting around the nucleus are the **electrons**. An electron has an equal and opposite electrical charge to a proton. A neutral (uncharged) atom has equal numbers of protons and electrons so that their opposite charges cancel out. The electrons are held within the atom by the electrical attraction of the nucleus. If an atom gains or loses electrons it becomes a charged particle, an **ion**.

**WHY IT MATTERS**
Different elements have different atomic structures

**KEY THINKERS**
John Dalton
(1766–1844)
J. J. Thomson
(1856–1940)
Ernest Rutherford
(1871–1937)
Maria Goeppert-Mayer
(1906–1972)

**WHAT COMES NEXT**
The periodic table

**SEE ALSO**
Radioactive decay,
p.114

# Atomic structure

By the turn of the 20th century most scientists had accepted that matter is made up of atoms. These combine in various ways to make molecules of all the different materials we see around us. But what is the structure of an atom?

Every atom has a central nucleus made up of two sorts of particles: protons and neutrons. Around the **nucleus** is a cloud of orbiting **electrons.** That's fairly easy to picture, although **quantum mechanics** suggests that the electrons in an atom should be thought of as a cloud of negative electric charge around the nucleus.

Since most of the mass of the atom is possessed by the nucleus and the nucleus is only a tiny part of the volume of the atom, it follows that the nucleus is very dense – many trillions of times the **density** of ordinary matter – and most of the atom is empty space.

# 56

# Electrons

**WHY IT MATTERS**
Electrons are
elementary particles

**KEY THINKERS**
J. J. Thomson
(1856–1940)
Robert Miilikan
(1868–1953)
Wolfgang Pauli
(1900–1958)

**WHAT COMES NEXT**
Electrons explain
chemical bonding

**SEE ALSO**
Structures of
materials, p.16

**Electrons** are negatively charged particles that orbit the **nucleus** of an atom. It is relatively easy to remove the outermost electrons from an atom. This is what happens when you pull a plastic comb through your hair and static electricity makes your hair stand on end – electrons have been stripped from some of the atoms of your hair and transferred to the comb. The strands of hair are now positively charged and so they repel each other (see page 62).

Something similar happens when atoms join together to form molecules. An electron is transferred from one atom to another, attracted by the positive charge of its nucleus. This forms a **bond**, holding the atoms together.

We know a lot about electrons. They have a tiny **mass**, about 1/1,840 of the mass of a proton or neutron. Their charge seems to be exactly the same as that of a proton, but opposite in sign – no one knows yet why they are exactly the same magnitude. And no one knows how big an electron is – measurements have determined that its radius is less than one tenth of a sextillionth of a metre, and it may have no size at all (zero radius).

All **electrons** are identical to one another and as far as we know they are **elementary particles** – they are not made of smaller particles. Their negative electric charge exactly balances the positive charge of protons. Their size is known to be less than one-billion-billionth of a metre. Because electrons are towards the outside of atoms they can move relatively easily from one atom to another. This is the basis of the chemical bonding of atoms to form molecules. In metals, the outermost electrons of atoms can move throughout the material, making metals good **conductors** of electric current and heat.

## 57

# The periodic table

**WHY IT MATTERS**
The periodic table organises the known elements into a logical pattern

**KEY THINKERS**
Dmitri Mendeleev
(1834–1907)

**WHAT COMES NEXT**
The table helps to predict how atoms will behave chemically

**SEE ALSO**
Chemistry and biochemistry, p.18

The periodic table of the **elements** was developed by Dmitri Mendeleev in the late 1860s. It is a thing of (scientific) beauty. It presents all the known elements in a logical order. It starts with hydrogen, the lightest element, each of whose atoms has a single proton in its **nucleus**. Next is helium with 2 protons in its atomic nucleus (its **atomic number** is 2), and so on, up to the elements with the most protons. After uranium (atomic number 92), the elements are artificially made because their atoms are radioactive and decay rapidly.

The numbers of neutrons also increase from one element to the next but not in such an orderly way.

There's more to the table than this. It shows up patterns in the physical and chemical properties of the elements. For example, elements that are gases at room temperature are close to the top corners of the table; most of the elements in the centre of the table are metals.

**in 100 words**

An **element** is a pure substance, each atom of which has the same number of protons in its **nucleus**. In the periodic table, the elements are arranged in increasing order of number of protons (their **atomic number**). Vertical columns are known as groups and contain elements with similar chemical properties. Horizontal rows are known as periods; the physical and chemical properties of elements vary in a regular way across each period.

You may come across different versions of the table, organised or annotated to emphasise different aspects. However, the elements will always be arranged in order of increasing atomic number.

# 58

# Isotopes

**WHY IT MATTERS**
The atoms of elements
exist in different
physical forms

**KEY THINKERS**
Frederick Soddy
(1877–1956)
J. J. Thomson
(1856–1940)
Francis Aston
(1877–1945)

**WHAT COMES NEXT**
Some isotopes are
more stable than
others

**SEE ALSO**
Radioactive
decay, p.114

The **isotopes** of an element are different atomic forms of that **element**. In everyday speech the word "isotope" has come to have a rather sinister sense, because we hear about hazardous isotopes in the news. Strontium-90 spread around the world through nuclear weapons testing. Polonium-210 was used to murder Alexander Litvinenko, a former Russian agent; the waste from nuclear power stations contains many different hazardous isotopes. Some of these are hazardous because they are radioactive, others because they are poisonous.

In fact, every atom is an isotope of its element. Take water, $H_2O$. Each hydrogen atom (H) has one proton and no neutrons in its **nucleus**. This is hydrogen-1. But a small fraction of the hydrogen atoms will have one proton and one neutron in their nuclei. This is a different isotope of hydrogen, hydrogen-2. All oxygen atoms have 8 protons in their nuclei; most will have 8 neutrons but some will have 9 or 10 neutrons. That gives three isotopes: oxygen-16, -17 and -18.

# In
# 100
## words

The atoms of an **element** all have the
same number of protons in their **nucleus**,
but they may have different numbers of neutrons.
These different forms are known as isotopes
of the element. The isotopes of an element have very similar
chemical properties (because they have the same numbers of
protons and **electrons**). However, some will be stable while
others will be unstable and undergo radioactive decay
(see page 114).

Each isotope is defined by two numbers.
For example, uranium-235 has 92
protons (as do all isotopes of
uranium); it has 235 − 92
= 143 neutrons
in its nucleus.

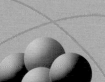

Representations of two
isotopes of helium, He:
helium 3 has 2 protons
and 1 neutron in its
nucleus; helium-4 has
one more neutron

$^{3}_{2}$ He

$^{4}_{2}$ He

# 59

# Nuclear forces

We know that **electrons** tend to stick to their atoms because their negative electric charge is attracted to the positive charge of the protons in the **nucleus**. But why doesn't the nucleus blast itself apart? All those protons must repel each other because they all have positive charge.

Another force acts to hold the nucleus together, the strong **nuclear force**. It attracts protons and neutrons together, but only when they are close together. In large nuclei – for example, gold – there are more neutrons than protons; they are needed to provide extra "glue" to hold the nucleus together against the electrical repulsion between the large numbers of protons.

in
**100**
words

The strong nuclear force is an attractive force that acts on protons and neutrons. In a stable (non-radioactive) **nucleus**, this force balances the electrical repulsion between protons. In most nuclei, neutrons help to reduce the repulsion between protons. A second nuclear force, the weak nuclear force, acts within nuclei and is involved in radioactive decay. These two nuclear forces are among the four fundamental interactions that give rise to all the forces between objects. The other two are the electromagnetic force and gravitation. The two nuclear forces are described as "short range" as they do not extend beyond the radius of a proton.

# 60

# Radioactive decay

In 1896 Henri Becquerel was studying rocks containing uranium. They were phosphorescent – they glowed in the dark. Initially he thought they must absorb light during the day and emit it later, but no matter how long he kept his rocks in the dark they continued to glow. They seemed to have an invisible energy source with no way to switch it off.

Uranium is a radioactive **element**. All of its **isotopes** have unstable nuclei. They gradually decay to become nuclei of other elements, emitting radiation as they do so. A Geiger counter can detect the radiation coming from a radioactive substance. It's amazing to think that each click of the counter means that it has detected the decay of a single tiny atomic **nucleus**.

Another feature of the clicks of a Geiger counter is that they are unpredictable – they come at random. There is no way of knowing when an individual atomic nucleus will decay. This randomness is why we talk about the half-life of a radioactive substance. Although we can't predict when an individual nucleus will emit radiation, we can measure how long it takes on average for half a sample to decay.

There are about 250 **isotopes** known to be stable. More than 3,000 are unstable – they are radioactive. The forces between the protons and neutrons are not in perfect balance for all time. In the periodic table, no **element** beyond lead (**atomic number** 82) has a stable isotope. In radioactive decay, an unstable **nucleus** emits radiation and becomes an isotope of a different element. The process of radioactive decay is unpredictable, so radiation is emitted randomly over time. To compare different isotopes we measure their half-lives, the average time taken for half the nuclei in a sample to decay.

# 61

# Radiation

**WHY IT MATTERS**
Understanding radioactivity helps protect us from its hazards

**KEY THINKERS**
Ernest Rutherford (1871–1937)
Marie Curie (1867–1934)
Frederick Soddy (1877–1956)

**WHAT COMES NEXT**
Radiation has many uses as well as being hazardous

**SEE ALSO**
X-rays and gamma rays, p.52

The radiation hazard symbol, known as a trefoil, has three triangles radiating out from a central disc. These are designed to remind you of the three types of radiation that may be emitted by a radioactive substance – alpha, beta and gamma radiation. The first two are particles, the third is a form of electromagnetic radiation (see page 50).

All can be hazardous to humans. Alpha radiation cannot penetrate our skin, but if an alpha source gets inside you – for example, if you breathe in a radioactive gas – its radiation can be highly damaging to cells and can lead to cancer. Beta and gamma are more penetrating but less hazardous.

Physicists discovered that beta radiation is fast-moving **electrons**, and when these are emitted from radioactive nuclei another particle is emitted – an antineutrino, a form of **antimatter** (see page 120).

"My experiments proved that the radiation of uranium compounds ... is an atomic property of the element of uranium. Its intensity is proportional to the quantity of uranium contained in the compound, and depends neither on conditions of chemical combination, nor on external circumstances, such as light or temperature."

**Marie Curie**, double Nobel prize winner

Alpha radiation consists of particles that each consist of two protons and two neutrons – these are the same as the nuclei of atoms of helium-4: they are slower but much more massive than beta radiation and hence can be more damaging to living tissue.

Beta radiation is energetic **electrons**. In beta-decay, a neutron in a **nucleus** emits an electron and becomes a proton; an antineutrino is also emitted. In alpha and beta decay, the number of protons in a nucleus changes. It has become the nucleus of a different **element**. Gamma rays may be emitted alongside alpha or beta particles.

# 62

# Fission and fusion

**WHY IT MATTERS**
Fission can be
controlled to release
energy in nuclear
power stations

**KEY THINKERS**
Ida Noddack
(1896–1978)
Lise Meitner
(1878–1968)
Otto Hahn
(1879–1968)
Otto Frisch
(1904–1979)

**WHAT COMES NEXT**
One day it may be
possible to use fusion
in power stations

**SEE ALSO**
The life of a star, p.157

Austrian-born physicist Lise Meitner was one of many Jewish scientists who left Germany in the 1930s at the time of the rise of the Nazis. With her nephew Otto Frisch she continued her work on nuclear physics in Sweden, where they discovered **nuclear fission**. The nuclei of heavy **elements** such as uranium are unstable and decay radioactively – but Meitner showed that there was another possibility. Rarely, an unstable heavy **nucleus** could split into two parts, plus two or three spare neutrons. This is spontaneous fission.

Stimulated fission is another form of fission, used in nuclear power reactors and in atom bombs. A beam of neutrons is directed into uranium. If a nucleus absorbs a neutron it becomes more unstable and splits. This releases more neutrons which go on to stimulate more nuclei to split, and so on. Energy is released. For each gram of starting material, the energy released is millions of times greater than when fossil fuels are burned.

In **nuclear fusion** light nuclei are forced together (to overcome the electrical repulsion between their protons) so that they merge to become a single heavier nucleus. In both fusion and fission the total **mass** of the particles is less after the event; the decrease in mass can be used to calculate the energy released using $E = mc^2$ (see page 102).

In nuclear fission a heavy **nucleus** splits to form two lighter nuclei. These are more stable than the original nucleus because their protons and neutrons are more tightly bound together. Energy and neutrons are released in the process. If a neutron goes on to cause another fission event a chain reaction may be established.

In nuclear fusion two light nuclei are forced to merge. This is an unlikely event because of the electrical repulsion of each nucleus on the other. But at very high temperatures, such as exist inside stars, nuclei are moving so fast that they may overcome this repulsion.

# 63

# Antimatter

A cancer patient may have a PET scan to investigate their tumour. They are injected with a radioactive substance designed to accumulate at points of interest in her body. As it decays it emits gamma radiation, which can be detected outside the body and used to give an image of the organs of concern.

PET stands for Positron Emission Tomography. The radioisotopes used, usually fluorine-18 or oxygen-18, are emitters of **positrons**, an **antimatter** form of the **electron**. When a positron collides with an electron in the patient's body they annihilate each other. All that is left is two gamma rays, which emerge from the patient's body in opposite directions. By tracking these gamma rays backwards it is possible to determine exactly where in the body they were produced.

Scientists have proposed that antimatter could be a fuel for future space flights beyond the solar system, but it has proved exceedingly difficult to make and store significant quantities.

Each particle that makes up matter has a "mirror-image" antiparticle whose properties are opposite to it. For example, a **positron** is an **antimatter electron** with the same **mass** but opposite charge. When a particle meets its antiparticle they annihilate, leaving nothing but electromagnetic radiation. The energy released can be calculated using $E = mc^2$ (see page 102). The process can be reversed; a photon of electromagnetic radiation can be transformed into a particle and its antiparticle. This suggests that there should be equal amounts of matter and antimatter in the universe—no one understands yet why matter dominates.

"If an alien lands on your front lawn and extends an appendage as a gesture of greeting, before you get friendly, toss it an eightball. If the appendage explodes, then the alien was probably made of antimatter. If not, then you can proceed to take it to your leader."

**Neil deGrasse Tyson**, astronomer

## 64

# Quarks and leptons

**WHY IT MATTERS**
The standard model organises the elementary particles and their interactions into a logical pattern

**KEY THINKERS**
Julian Schwinger (1918–1994)
Richard Feynman (1918–1988)
Abdus Salam (1926–1996)
Peter Higgs (1929– )
Sau Lan Wu (c.1940–)

**WHAT COMES NEXT**
Gravity, dark matter and dark energy have yet to be incorporated into the standard model

**SEE ALSO**
Dark matter, p.166

It's hard finding out about the particles of which matter is made. The biggest molecules such as proteins can be seen using **electron** microscopes and individual atoms may be revealed using a scanning tunnelling microscope. But the **nucleus** of an atom is 100,000 times smaller than an atom, as are the protons and neutrons of which it is made. How can we find out what such small particles are made of? And what of some of the other particles of matter which were revealed by studies of cosmic rays (see page 124)?

Much of what we know has come from experiments in particle accelerators, also known as atom smashers (see page 70). Highly energetic particles are made to collide and, in the resulting debris, subatomic particles can be detected. Some have much greater **mass** than the original particles – energy has become mass.

The **standard model** summarises the known **elementary particles** of matter. Protons and neutrons are each made of three **quarks**, so it is quarks rather than protons and neutrons that are elementary particles. There are six types of quark, which can combine in twos and threes to make a variety of heavier particles. Alongside these are the **leptons**, which include **electrons** and neutrinos. Each of these comes in three "generations", making six in all. The standard model also includes three **gauge bosons**, particles that explain three of the four fundamental forces (excluding gravity), and the **Higgs boson** which explains **mass**.

The particles of the standard model: there are six quarks and six leptons, each with their antiparticles, plus five bosons which carry the forces between particles

| | 1st | 2nd | 3rd | | |
|---|---|---|---|---|---|
| **QUARKS** | $u$ up | $c$ charm | $t$ top | $\gamma$ photon | $H$ Higgs boson |
| | $d$ down | $s$ strange | $b$ beauty | $W^{\pm}$ W boson | |
| **LEPTONS** | $e$ electron | $\mu$ muon | $\tau$ tau | $Z^0$ Z boson | **GAUGE BOSONS** |
| | $V_e$ neutrino electron | $V_\mu$ neutrino muon | $V_\tau$ neutrino tau | $g$ gluon | |

# 65

# Cosmic rays

The Apollo programme astronauts travelled beyond the protection of the Earth's **magnetosphere**. They reported seeing occasional flashes of light that were thought to be caused by cosmic rays, particles travelling through space with enough energy to penetrate their spacecraft and their skulls. It's not clear whether these flashes resulted from the cosmic rays passing through their eyeballs or if the rays triggered an effect in the optic nerve or in brain cells.

Cosmic rays are likely to be a major problem for space flights to other planets or beyond the solar system. As well as affecting astronauts they can damage the electronic circuitry needed to control a spacecraft. Engineers may need to design magnetic shielding to deflect the rays.

**WHY IT MATTERS**
Cosmic rays reveal the nature of matter at high energies

**KEY THINKERS**
Pierre Auger
(1899–1993)
Bruno Rossi
(1905–1993)
Marietta Blau
(1894–1970)
James Cronin
(1931–2016)

**WHAT COMES NEXT**
The most energetic cosmic rays may tell us about interactions in distant galaxies

**SEE ALSO**
Space travel, p.148

### In 100 words

Cosmic rays are high-energy particles (mostly protons or atomic nuclei) travelling through space. They may come from the Sun, from elsewhere in our galaxy, or even from other galaxies. Exposure to cosmic rays is hazardous for astronauts and for aircrew in high-flying aircraft.
A cosmic ray reaching the Earth's atmosphere may be deflected by the Earth's **magnetosphere** or collide with an atmospheric molecule and produce a shower of secondary rays. These include gamma rays and particles of matter and **antimatter**. Capturing these has revealed particles such as muons and pions that are not familiar from ground-based experiments.

# Quantum Mechanics

Up to the turn of the 20th century, classical physics had increasing success in explaining the world around us. Some physicists even declared that they understood everything and that applying the known laws would solve all outstanding problems. But once their fellow scientists developed techniques to probe matter on the atomic scale, they saw that at that level the known laws were inadequate.
Quantum physics was born.

The word "quantum" means a quantity or amount. The idea in quantum physics is that when two objects interact, energy is parcelled up in fixed quantities. In-between quantities are not allowed. Classical physics says that any value of energy is possible; quantum physics says no. It's a bit like money; the smallest allowed quantity is one penny or one cent. Quantum physics holds many surprises. In particular, we can often only deduce the probabilities of things happening, rather than finding a definite outcome. This haziness or uncertainty is a characteristic of how things are on the atomic scale.

# 66

# Photons

**WHY IT MATTERS**
Light can be thought of
as being made up of
tiny particles called
photons

**KEY THINKERS**
Albert Einstein
(1879–1955)
Max Planck
(1858–1947)

**WHAT COMES NEXT**
The idea of photons
helps to explain the
structure of atoms

**SEE ALSO**
Quarks and
leptons, p.122

If you have an X-ray at the dentist's surgery, the dentist will stand well back or leave the room as you are exposed to the X-rays that are being used to show up the insides of your teeth. X-rays are hazardous to life – pregnant women are rarely X-rayed because of the danger to their foetus.

So what makes X-rays more hazardous than other types of **electromagnetic wave**, such as visible light or infrared radiation (see page 50)? To answer this we have to imagine these types of radiation travelling as photons rather than as waves. A photon is a particle of electromagnetic radiation and the photons that make up X-rays have much more energy than those of visible light or infrared.

Because X-ray photons have a great deal of energy they can do a lot of damage as they pass through living tissue. In particular they can damage the DNA in cells, causing them to divide uncontrollably – a cancerous tumour has been formed. For medical purposes the doses received by patients are unlikely to have this effect, but anyone working with X-rays must be careful to avoid regular exposure.

Electromagnetic radiation can be thought of as travelling in the form of photons. A photon is a **quantum** of energy, a massless particle that travels at the speed of light. A photon emitted by one atom will travel until it is absorbed by another atom – it disappears as its energy is transferred to the second atom. The energy of a photon of radiation is related to the frequency of the radiation; higher-energy photons correspond to light of a higher frequency. Since X-rays and gamma rays have the highest frequencies in the **electromagnetic spectrum** their photons have the most energy.

# 67

# The exclusion principle

**WHY IT MATTERS**
This principle governs the energy values of electrons in an atom

**KEY THINKERS**
Wolfgang Pauli (1900–1955)
Niels Bohr (1885–1962)

**WHAT COMES NEXT**
The exclusion principle helps to explain which frequencies of light are absorbed or emitted by atoms

**SEE ALSO**
Atomic structure, p.105

Atoms are often represented as having a central **nucleus** with **electrons** orbiting around it. People sometimes use this image simply to suggest that a scientific subject is being discussed.

The image is correct in that it shows the electrons going round and round; this idea was a surprise to physicists trying to explain the structure of the atom because the laws of electromagnetism suggested that each electron should rapidly lose energy (in the form of electromagnetic radiation – see page 50) and spiral in towards the nucleus.

The fact that this doesn't happen fits with the idea that electrons in an atom can only have certain values of energy. If they lost a little bit of energy they would no longer have an allowed value of energy.

Something else that this image doesn't show is that all the electrons in an atom must have different energies. Within an individual atom, each electron occupies a different energy level. This idea was developed by Wolfgang Pauli and is known as Pauli's exclusion principle.

In
**100**
words

**Electrons** in an atom can only have
certain fixed values of energy – their energy
is **quantised**. These allowed energy levels are
like a ladder with irregularly spaced rungs.
Within an atom, no two electrons may occupy the same
energy level. This is Pauli's exclusion principle. In a metal, electrons
from many atoms interact with each other and so they
occupy many closely spaced energy levels.
When a material is cooled towards **absolute zero**
(see page 96), the electrons
occupy only the lowest energy levels.
At higher temperatures they
start to have enough
energy to rise
to higher
levels.

# 68

# Electrons and photons

Fireworks light up the sky with their varied colours. These different colours come from the different chemical **elements** that are burning and giving out light. Strontium gives red, sodium gives yellow, barium gives green, for example.

Studying the spectrum of light from a burning substance shows that only a few frequencies of light are present, different frequencies for different elements. That's useful, making it possible to identify the elements that are present in a sample by looking at the frequencies of light emitted when it burns. Forensic scientists use this when examining samples from a crime scene.

**WHY IT MATTERS**
Analysing the light from atoms of different elements reveals that their electrons are confined to discrete energy levels

**KEY THINKERS**
Janne Rydberg (1854–1919)
Niels Bohr (1885–1962)

**WHAT COMES NEXT**
New rules must be developed to explain how electrons behave in atoms

**SEE ALSO**
Spectra, p.48

## In 100 words

An atom emits light when one of its **electrons** changes its energy. The electron loses energy and a single photon of light is emitted, carrying off the energy lost by the electron. A bigger change in energy gives a more energetic, higher-frequency photon. Similarly, an electron can jump to a higher, empty level if it absorbs a photon of energy equal to the difference between the two levels. Atoms of different elements have differently spaced energy levels and so emit photons with different frequencies. The frequencies of light in an element's spectrum reveal the pattern of energy levels.

## 69

## In 100 words

Albert Einstein explained the **photoelectric effect** in 1905. A photon of light strikes a metal surface where many **electrons** are free to move around. The photon is absorbed by a single electron, giving it enough energy to escape from the metal. If the frequency of the light is too low its photons have insufficient energy to free an electron from the metal. Making the light more intense increases the numbers of photons arriving each second and so electrons are released at a greater rate. The effect is used in devices including camera light meters, photomultipliers and night-vision cameras.

**WHY IT MATTERS**
The photoelectric effect reveals how light interacts with matter

**KEY THINKERS**
Pythagoras and Hippasus
(c. 600–400BCE)
Euclid (c.325–265BCE)

**WHAT COMES NEXT**
Dust rises from the Moon's surface as a result of the photoelectric effect

**SEE ALSO**
Wave particle duality, p.132

# Photoelectric effect

The light coming from distant stars is very faint. Astronomers use photomultipliers to catch every photon of light entering their telescopes. In a photomultiplier, light falls on a metallic surface. Each photon knocks out a single **electron** – in effect, a tiny electric current starts to flow. This is the **photoelectric effect**. The current can be amplified up to 100 million times and contributes to the image of the star being observed.

The photoelectric effect was observed in the 19th century but it was impossible to explain in terms of light waves. Only by thinking of light as individual photons, each releasing a single electron from the metal, could the effect be understood.

# 70

# Wave particle duality

**WHY IT MATTERS**
This idea reveals the underlying nature of matter and energy

**KEY THINKERS**
Max Planck
(1858–1947)
Louis de Broglie
(1892–1987)
Erwin Schrödinger
(1887–1961)

**WHAT COMES NEXT**
Particles can be represented mathematically as waves

**SEE ALSO**
Quantum interpretations, p.138

**Electron** microscopes are powerful devices that can reveal much smaller details than a conventional light microscope. The problem with light microscopes is that visible light has a wavelength a little less than one-thousandth of a millimetre. If light shines on anything smaller than this it will **diffract** (see page 44) and the image of the object becomes blurred.

It's surprising to find that a beam of electrons can also be diffracted – that's a phenomenon associated with waves. However, a beam of fast-moving electrons can have a wavelength thousands of times smaller than that of visible light and so can reveal objects as small as protein molecules that could never be seen using a light microscope.

"Electron waves" passing through regular arrays of atoms or molecules produce **interference** patterns that reveal the arrangements of those particles. This makes electron diffraction a powerful tool for revealing the structures of materials.

The **photoelectric effect** (see page 131) can be explained by thinking of light as photons – that is, as **quanta** of energy interacting with **electrons** that behave as particles. But to explain electron **diffraction** we have to imagine that electrons travel as waves, like light waves, and show wave effects such as **interference**.

It appears that everything we think of as particles can have a wave nature, and every wave has a particle nature. This is known as **wave particle duality** – in explaining physical phenomena, it is necessary to choose one or other of these descriptions of particles and waves.

# 71

# Quantum tunnelling

**WHY IT MATTERS**
Tunnelling explains
phenomena that
classical physics
cannot

**KEY THINKERS**
George Gamow
(1904–1968)
Max Born (1882–1970)

**WHAT COMES NEXT**
Tunnelling is used in
many electronic
devices

**SEE ALSO**
Quantum
uncertainty, p.136

Imagine that you are inside a room with no door to the outside world. You rush around, colliding with the walls over and over again, but you're still stuck in the room. Now imagine that on one lucky occasion you crash into the wall and suddenly find yourself outside. The wall remains intact. Impossible?

This *is* impossible on the scale of human beings and the large-scale world we inhabit, but remarkably it is possible on the atomic scale. A particle may not have enough energy to escape from inside an atomic **nucleus** but there is a tiny chance that it may still get out. That's how radioactive decay happens (see page 114). This effect is known as **quantum** tunnelling.

Tunnelling helps to explain **nuclear fusion** (see page 118) where two light atomic nuclei merge to form a heavier nucleus. In principle the two nuclei should be kept apart by the electrical force with which they repel each other. They do not have enough energy to get close enough together for the nuclear strong force to act. However, there is just a tiny chance that they can tunnel through this "energy barrier", and this is what happens from time to time.

In **quantum mechanics** a particle can be thought of as wave-like and represented mathematically as a probability wave (its wave function) which tells us the probability that it is at a particular point in space. If a particle has insufficient energy to surmount an energy barrier, there is still a probability that its wave function is non-zero on the other side of the barrier. This means there is a chance that the particle will appear on the other side, having tunnelled through the barrier. Quantum tunnelling helps to explain radioactive decay and is used in scanning tunnelling microscopes.

## 72

# Quantum uncertainty

**WHY IT MATTERS**
At the quantum level
there are limits to
measurements of
physical quantities

**KEY THINKERS**
Louis de Broglie
(1892–1987)
Werner Heisenberg
(1901–1976)

**WHAT COMES NEXT**
Philosophers and
physicists debate the
nature of reality

**SEE ALSO**
Quantum
interpretations, p.138

When a radar speed trap is used to catch speeding motorists, a radar gun measures a car's speed and at the same time a photograph is taken to show its position. A slow camera will give a blurry image, but using a camera with a faster shutter speed will give a clearer picture so that the position of the car is known more precisely.

It might seem that, by using better and better equipment, the car's speed and position can be known more and more precisely. However, according to **quantum mechanics**, this is not so. Measuring an object's speed with great precision reduces the precision with which its position can be known (and vice versa). This has no effect for macroscopic situations like a speed trap, but it is important on the scale of atoms.

"Experiments designed to detect particles always detect particles; experiments designed to detect waves always detect waves. No experiment shows the electron behaving like a wave and a particle at the same time."

**John Gribbin,** author of *In Search of Schrödinger's Cat*

Werner Heisenberg's uncertainty principle says that it is impossible to know with perfect accuracy both an object's position and its speed (or, more correctly, its **momentum** – see page 28). Making the measurement of one quantity more precise reduces the precision with which the other can be measured. This applies to measurements of energy and the time of measurement.

**Quantum mechanics** describes the state of a particle using a mathematical formula called a wave function that shows the particle as a wave of probability spread over a region of space. This spread of probabilities leads to inherent uncertainties in measurements.

# 73

# Quantum interpretations

**WHY IT MATTERS**
It is difficult to relate quantum theory to our everyday experience

**KEY THINKERS**
Erwin Schrödinger (1887–1961)
Niels Bohr (1885–1962)
Hugh Everett (1930–1982)

**WHAT COMES NEXT**
Different ways of making sense of quantum theory are still debated

**SEE ALSO**
Mathematics, p.180

"Schrödinger's Cat" is a famous thought experiment devised to test different ways of interpreting the results of **quantum** theory. A cat is placed in a steel box with a flask of cyanide and a radioactive substance. There is a 50:50 chance of detecting a radioactive decay in an hour. If a decay is detected, the flask is cracked open, the cyanide is released and the cat dies. What will the situation be after an hour?

We can't see inside the box. **Quantum mechanics** says that there will be equal quantities of live cat and dead cat (because of the 50:50 probability). But it's only when we open the box that we will know if the cat is alive or dead. The mixture of live and dead cat is called a *superposition of states*; opening the box causes the superposition to collapse into one state or the other. Erwin Schrödinger developed this idea because he wasn't convinced that reality only comes into existence when we observe it, a version of quantum mechanics called the Copenhagen Interpretation after the city where it was developed by Niels Bohr and Werner Heisenberg.

An alternative approach is the Many Worlds interpretation, devised by Hugh Everett. This suggests that the outcome of the experiment will be two universes. In one the cat remains alive, in the other it is dead.

**Quantum** theory has been successful in enabling physicists to calculate the outcomes of interactions on the atomic scale. It represents objects such as atoms and **electrons** using mathematical formulae and uses these to deduce the outcomes of interactions. But many of its ideas are counter-intuitive and scientists have devised several different interpretations. In the Copenhagen Interpretation, reality only comes into existence when a system is observed. In the Many Worlds Interpretation, whenever there are two possible outcomes of an interaction, both come into existence and each continues into the future. These and other interpretations are debated by physicists and philosophers.

"Those who are not shocked when they first come across quantum theory cannot possibly have understood it."
**Niels Bohr,** physicist

# 74

# Super-conductivity

## WHY IT MATTERS
An electric current can flow forever in a superconductor without losing energy

## KEY THINKERS
Heike Kamerlingh Onnes (1852–1926)
Leon Cooper (1930–)
Brian Josephson (1940–)

## WHAT COMES NEXT
Physicists are searching for materials that are superconductors at room temperature

## SEE ALSO
Electric current, p.66

In a Magnetic Resonance Imaging (MRI) scanner, the patient lies with the part of their body to be scanned in the middle of a giant cylindrical electromagnet (see page 68). This magnet is a coil of superconducting wire cooled using liquid helium to a low temperature at which the wire has no electrical resistance. The magnetic field produced is strong, up to one million times the Earth's magnetic field.

An electromagnet made with ordinary wire has electrical resistance. This is because **electrons** moving through it collide with any irregularities such as impurity atoms or vibrating atoms and in the process lose energy to the material of the wire – it becomes hot. A conventional electromagnet is unsuitable for an MRI scanner as it would become exceedingly hot and be in danger of melting. Using a superconductor overcomes this problem.

Similar electromagnets are used to change the direction of particle beams in accelerators such as those at CERN (see page 70).

**In 100 words**

When there is an electric current in a metal, the moving **electrons** lose energy as they collide with vibrating atoms or with impurity atoms. This is the cause of electrical resistance and results in the metal becoming hot.

A superconductor is a material that loses all electrical resistance when it is cooled below a critical temperature, usually below 100K (-173°C/-279°F), an effect discovered in 1911 by Heike Kamerlingh Onnes.

**Quantum mechanics** describes how the electrons in the material bind together in pairs, reducing their energy so that they can no longer be scattered by irregularities in the material.

# 75 Lasers

The bright red beam that is used in supermarkets to scan barcodes is light from a laser. Although laser light is perfectly ordinary light, it has an unusual feature – it is all of one frequency (see page 39). That's very different from a traditional lightbulb, which has a hot wire filament and produces light with frequencies from across the visible spectrum.

"Laser" is an acronym of Light Amplification by Stimulated Emission of Radiation. Lasers have found so many uses that there are roughly ten times as many lasers as people on the planet.

**WHY IT MATTERS**
Lasers produce narrow beams of light of a single frequency

**KEY THINKERS**
Charles H Townes
(1915–2015)
Arthur Schawlow
(1921–1999)
Theodore Maiman
(1927–2007)

**WHAT COMES NEXT**
Lasers are used for making holograms

**SEE ALSO**
Spectra, p.48

**In**
## 100
**words**

A laser consists of a solid rod or a gas-filled tube with mirrored ends. Electricity is used to give energy to atoms so that **electrons** rise to a higher energy level. These are stimulated to drop to a lower level by a passing photon of light of the correct frequency, and a second, identical photon is produced; now there are two photons that can each stimulate further production of photons, and so the process escalates to produce a beam of vast numbers of identical photons.
A laser beam is "coherent" – all the photons are in step with one another.

# 76

A **quantum** computer uses **qubits** that can exist in a state where the two values 0 and 1 are superposed and which only take on a specific value when a result is extracted from the machine. Interconnected qubits can process multiple calculations simultaneously, unlike a conventional machine. As more qubits are added to the system the rate at which they can process data increases exponentially.
The great speed of quantum computers is enabling them to replace conventional machines in solving complex problems. Since they have to operate at very low temperatures, they are ` unlikely to replace laptop and desktop machines.

**WHY IT MATTERS**
**Quantum** computers can solve complex problems quickly

**KEY THINKERS**
Paul Benioff
(1930–2022)
Richard Feynman
(1918–1988)
Peter Shor (1959–)
Michelle Simmons
(1967–)

**WHAT COMES NEXT**
Experimental machines will be replaced by standardised hardware

**SEE ALSO**
Superconductivity,
p.140

# Quantum computing

At the heart of a conventional computer are streams of electrical pulses, representing binary digits (either 0 or 1) that are processed using large numbers of transistors in a silicon chip processor.

A **quantum** computer is based on **qubits**, or quantum bits, derived from **electrons**, photons or **ions**. These are tiny so that they obey the rules of **quantum mechanics**. First results from quantum computers suggest that they will be able to solve problems so complex that even supercomputers fail to solve them – for example, how to route large fleets of cargo vessels most efficiently, or how to analyse vast datasets from Earth-observation satellites (see page 151).

# Astronomy

In past millennia people had a better view of
the stars and planets and built them into their
myth-making and into their calendar of the
annual cycle of the seasons. Now, more than half
of the world's population lives in cities, where
light pollution makes the night skies difficult
to observe.

The last four centuries have seen great
developments in telescopes (see page 150). This
has allowed us to see far greater distances into
space and to see in much greater detail the
objects that exist in space. While the images we
see are fascinating in themselves, physicists have
had to work hard to explain their observations.
The ideas used to explain what's going on "out
there" must be compatible with the ideas we use
to explain what we observe down here in our
laboratories on the surface of the Earth.

The Sun's apparent daily motion across the sky from east to west is a result of the Earth's rotation from west to east, giving rise to night and day. As the Earth is also travelling around the Sun, the stars we see in the night sky are different at different times of year.

The picture of the solar system with the Sun at its centre and with the Earth and the other planets orbiting it is the heliocentric model. A convincing mathematical explanation was first published by Copernicus in 1543. The (incorrect) geocentric model has the Earth at the centre.

**WHY IT MATTERS**
Our idea of our place in the universe was changed forever

**KEY THINKERS**
Aristarchus
(310–230BCE)
Nicolaus Copernicus
(1473–1543)
Tycho Brahe
(1546–1601)

**WHAT COMES NEXT**
Understanding our galaxy

**SEE ALSO**
Orbits, p.147
Remote sensing, p.151

# The heliocentric model

We observe the Sun rising in the East, travelling across the sky and setting in the West. The stars do something similar, except that the pattern moves on slightly each night so that we see different sets of stars in the summer and the winter.

How do you account for this? An obvious explanation would be that these celestial objects are orbiting the Earth – it's hard to believe (though it's correct) that the Earth itself is rotating and that this causes the apparent motion of the Sun and stars across the sky. Although some ancient Greek thinkers had suggested that the Sun might be at the centre of things, rather than the Earth, it was another 2,000 years before this became the scientific orthodoxy in Europe, the Islamic world and India.

# 78

# Planets

The motion of the stars across the night sky is entirely predictable – the pattern is the same from one year to the next. But for a very long time people have known that five "stars" behave differently from all the other stars. These are the five brightest planets: Mercury, Venus, Mars, Jupiter and Saturn. Their positions change against the background of stars as they orbit the Sun.

William Herschel was the first astronomer to discover a previously unknown planet when he spotted Uranus in 1781. Pluto was discovered by Clyde Tombaugh in 1930; unfortunately for him it has since been downgraded and is classified as a minor planet.

**WHY IT MATTERS**
The planets' movement in the night sky is different from that of the stars

**KEY THINKERS**
Nicolaus Copernicus (1473–1543)
Johannes Kepler (1571–1630)
William Herschel (1738–1822)
Clyde Tombaugh (1906–1997)

**WHAT COMES NEXT**
Astronomers have identified eight planets and many hundreds of their moons

**SEE ALSO**
Exoplanets, p.152

In
**100**
words

The eight planets travel around the Sun – because of this, we see their positions against the background of stars changing slightly from night to night. The word "planet" means "wanderer" in ancient Greek. The planets' orbits lie roughly in a flat plane, the plane of the ecliptic. Consequently, an observer on the Earth sees the planets crossing the night sky along the line followed by the Sun during the daytime. Because Mercury and Venus are closer to the Sun than we are, we always see them close to the Sun. A transit is when either of these planets passes between us and the Sun.

## 79

## In 100 words

The planets are held in their orbits by the Sun's gravitational pull. Those furthest from the Sun, where its **gravitational field** is weakest, travel most slowly, so the period of their orbit (their "year") is longest. The distance of a comet from the Sun changes during its orbit, so its speed also changes. The planets' motion was successfully explained using Newton's laws of motion and his inverse square law of universal gravitation. But a detailed analysis of the orbit of Mercury showed that Newton's laws must be modified according to Einstein's general theory of relativity (see page 78).

# Orbits

Caroline Herschel became so familiar with the pattern of stars seen through a large telescope that she was quick to notice anything unusual. In this way she discovered eight comets between 1786 and 1797.

Comets follow elliptical orbits around the Sun. Most spend a long time far from the Sun, moving slowly. Then they gradually fall inwards, speeding up as they do so. Many fall into the Sun, but some swing around behind it and climb back out into the farthest reaches of the solar system.

Planets and comets are held in their orbits by the pull of the Sun's gravity (see page 72). To stay in their orbit they must travel at the right speed; too fast, and they will disappear out into space; too slow, and they will spiral down into the Sun.

**WHY IT MATTERS**
Physical laws allow us to predict the orbits of planets, comets and spacecraft

**KEY THINKERS**
Isaac Newton
(1643–1727)
Caroline Herschel
(1750–1848)
Albert Einstein
(1879–1955)
Katherine Johnson
(1918–2020)

**WHAT COMES NEXT**
Spacecraft have now landed on comets and asteroids

**SEE ALSO**
Remote sensing, p.151

# 80

# Space travel

**WHY IT MATTERS**
Spacecraft can reveal much more about the solar system and the stars beyond than Earth-based observation

**KEY THINKERS**
Konstantin Tsiolkovsky (1857–1935)
Robert H. Goddard (1882–1945)
Sergei Korolev (1907–1966)

**WHAT COMES NEXT**
Plans are underway to land people on Mars

**SEE ALSO**
Remote sensing, p.151

If you throw a stone vertically upwards it will slow down as it ascends through the Earth's **gravitational field**, and then speed up as it descends again. If you could throw a stone so that it leaves your hand at 11km/s (6.8 miles/s), it would move upwards forever – you would have given it enough energy to escape from the Earth's gravity. This speed is known as escape velocity.

It takes about 60 megajoules of energy to get 1kg out into space. That's more than the energy stored in 1kg of wood, which is why there are no wood-fired space rockets. (1kg is 2.2lb.)

For people to travel to another planet and back will require a lot of energy: energy to get free from the Earth, energy to slow down the spacecraft so that it doesn't crash on landing, and the same again to get back home. That's why scientists are looking for ways of manufacturing rocket fuel on a planet such as Mars using sunlight and other local resources.

A spacecraft orbiting the Earth travels at about 8km/s (5 miles/s). No fuel is required once in orbit since the spacecraft is held there by Earth's gravitational pull.
Similarly, no fuel is required once a spacecraft is free of Earth's gravity – it will travel at a steady speed in a straight line, according to Newton's first law of motion (see page 21). But to change speed or direction requires fuel for the rocket motors – or, in a slingshot manoeuvre, the spacecraft uses the gravitational pull of another planet to alter its speed and course.

"I never went into physics or the astronaut corps to become a role model. But after my first flight, it became clear to me that I was one. And I began to understand the importance of that to people."
**Sally Ride**, astronaut

# 81

# Telescopes

In 1609, when Galileo Galilei used an early optical (light) telescope to observe Jupiter, he noticed four bright "stars" lying along a line that crossed the planet. Their positions changed from night to night and from this he deduced that they were moons in orbit around Jupiter.

Galileo's discovery, and his studies of the Moon's cratered surface and the phases of Venus, effectively put paid to the idea that celestial objects such as the planets, Sun and stars were somehow perfect and made from entirely different stuff from Earth and its inhabitants. The evidence of telescopes could not be denied.

**WHY IT MATTERS**
Telescopes create images of distant objects

**KEY THINKERS**
Galileo Galilei
(1564–1642)
Percival Lowell
(1855–1916)
Karl Jansky
(1905–1950)
Ruby Payne-Scott
(1912–1981)

**WHAT COMES NEXT**
Space telescopes can look back in time towards the Big Bang

**SEE ALSO**
Refraction, p.46

## In 100 words

A telescope gathers **electromagnetic waves** arriving from space and constructs an image of the waves' source. The first telescopes had lenses and mirrors to collect visible light but today's telescopes cover all regions of the **electromagnetic spectrum** (see page 50).
The Earth's atmosphere absorbs all but radio waves, visible light and parts of the infrared and ultraviolet regions of the spectrum. Space telescopes in orbit above the atmosphere have greatly expanded our knowledge as they can gather X-rays, gamma rays and other waves that would otherwise be absorbed. Gravitational wave detectors are sometimes also called telescopes.

Remote sensing involves making all kinds of measurements without having direct contact with the target. Orbiting spacecraft can use radar to determine their height above the Earth's surface and so calculate changes in the height of the oceans' surface. Since trees, pastures, rocks and different soils reflect different parts of the spectrum of sunlight it is possible to produce maps showing land use and how it is changing. Weather satellites, which detect changes in temperature, pressure, wind speed and humidity, have allowed great improvements in weather forecasting. The vast amounts of data generated require great computing power for processing.

**WHY IT MATTERS**
Earth-observation satellites reveal the effects of climate change

**KEY THINKERS**
Konstantin Tsiolkovsky (1857–1935)
James Lovelock (1919–2022)

**WHAT COMES NEXT**
Remote-sensing spacecraft have visited other planets

**SEE ALSO**
Cosmic rays, p.124

# Remote sensing

While many telescopes are directed outwards to target astronomical objects, there are many satellites orbiting the Earth that have sensors directed downwards towards our planet's surface. They observe a great range of features – for example, the European Space Agency's Sentinel spacecraft uses radar to monitor the Amazon rainforest and record deforestation events.

# 83

# Exoplanets

Scientists have long imagined that the stars they see in the night sky must have planets orbiting them. Now we know that this is true.

In 1995 Michel Mayor and Didier Queloz reported the existence of a massive planet orbiting close to its parent star in the constellation of Pegasus. What they actually observed was variations in the spectrum of the light coming from the star. It varied periodically from being **red-shifted** to blue-shifted and back again, an example of the **Doppler effect** (see page 58). This variation repeated every four days, which they deduced to be the period of the orbiting planet.

Such a massive planet orbiting close to its star is called a "hot Jupiter" because it is a giant planet like Jupiter and its proximity to its star makes it hot.

An exoplanet or extrasolar planet is a planet orbiting a star other than the Sun. Astronomers have detected thousands, with some stars having multiple planets in orbit around them. Scientists use direct observation by telescope to find exoplanets. Another approach is to watch for a periodic decrease in the brightness of a star as an exoplanet orbits across the face of the star, while a third uses the way a star appears to wobble as it and its exoplanet orbit around their common centre of gravity.
Detection of oxygen in an exoplanet's atmosphere may hint at the presence of life.

One way to detect an exoplanet is to observe the slight dimming of its star's light as it passes across the star's disc

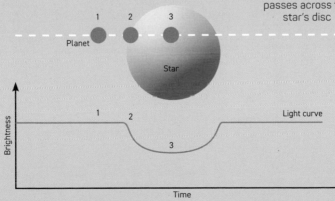

# 84

# The Fermi paradox

The Roswell Incident of 1947 never ceases to provoke debate. Some people claim debris found near Roswell, New Mexico, was that of a crashed alien spacecraft covered up by the authorities, but the US Army states that it was in fact the remains of a weather balloon. Many people believe they have seen unidentified flying objects (UFOs) and a good number believe they have been temporarily captured by aliens, taken into their spacecraft and examined. And yet few scientists would claim that there is any significant evidence of Earth having been visited by beings from a civilisation elsewhere "out there".

Since we know that there are many planets similar to Earth orbiting other stars in our galaxy, and that the solar system has been in existence for more than four billion years, shouldn't at least one alien civilisation have made contact by now? As Enrico Fermi reputedly said while debating the existence of alien civilisations with a group of colleagues, "Where are they?"

> "There may be aliens in our Milky Way galaxy, and there are billions of other galaxies. The probability is almost certain that there is life somewhere in space."
>
> **Buzz Aldrin**, astronaut

The Fermi paradox suggests that, given the many billions of stars in our galaxy and the billions of years which have elapsed since its formation, it is likely that intelligent life will have evolved on many occasions. Such life could be expected to have invented systems of interstellar travel capable of reaching Earth. But there is no reliable evidence that this has happened.

Among many explanations is the suggestion that alien civilisations are reluctant to reveal themselves, or that they are prone to self-destruction. Alternatively, they may be observing us now using advanced technologies that are beyond our detection.

# 85

# The Milky Way

The Milky Way is the band of stars crossing the night sky which, unfortunately, has become less visible to many observers because of light pollution. It has a long history of myth-making associated with it but, in 1610, Galileo used his telescope to determine that it is a **mass** of stars, a galaxy of which our star the Sun is just one.

From within the galaxy it is tricky to determine its shape – it's impossible to send a spacecraft far enough into space to see it from outside. However, given its appearance as a narrow band, we can say that it is a thin, somewhat two-dimensional, distribution of stars. Counting stars in different directions has enabled astronomers to show that the solar system is about one-third of the way out from the centre of a barred spiral galaxy.

**WHY IT MATTERS**
The Milky Way is our local environment in space

**KEY THINKERS**
Galileo Galilei
(1564–1642)
Heber Curtis
(1872–1942)
Edwin Hubble
(1889–1953)

**WHAT COMES NEXT**
The rotation of a galaxy is determined by the presence of dark matter

**SEE ALSO**
Dark matter, p.166

In
**100**
words

The Milky Way is a barred spiral galaxy made up of a few hundred billion stars. It is held together by the gravitational attraction (see page 72) between the masses of the stars and dark matter (see page 166).
The galaxy is much wider than it is thick so that, as we look outwards into its disc, we see it as a band of stars across the sky. Until about a century ago it was believed that the Milky Way contained all the stars in the universe, but now many billions of other galaxies have been detected.

86

**In 100 words**

Stars form by inward gravitational attraction of dust clouds. As this material falls inwards its gravitational **potential energy** (see page 86) changes to **kinetic energy**. Eventually the particles are moving fast enough for **nuclear fusion** to occur, releasing energy as light from the surface of the new star. For a long time the outward pressure of the fast-moving particles balances the inward pull of gravity. Eventually the star runs out of fuel and expands to become a cooler red giant. This may fizzle out as a white dwarf; for more massive stars, an energetic **supernova** explosion may occur.

# The life of a star

Stars form in regions of space where there are clouds of dust and gas. These materials collapse inwards, pulled together by the gravitational attraction of their **masses**. As they fall inwards they speed up so that their **kinetic energy** increases. As they collide they share their energy and so the material becomes hotter and hotter. Eventually it will become hot enough for **nuclear fusion** to start (see page 118) and it starts to glow – a star is born.

The Sun formed in this way almost five billion years ago. It will "burn" at a fairly steady rate for a few more billions of years but then it will blow up to become a red giant, engulfing the Earth. Later it will collapse inwards to become a white dwarf, and gradually fade away.

**WHY IT MATTERS**
Explains how stars form and later die

**KEY THINKERS**
Annie Jump Cannon (1863–1941)
Henrietta Leavitt (1868–1921)
Ejnar Hertzsprung (1873–1967)
Henry Norris Russell (1877–1957)

**WHAT COMES NEXT**
Space telescopes have revealed "nurseries" in which stars are forming

**SEE ALSO**

Nucleosynthesis, p.164

# Cosmology

For traditional societies the cosmos
represented an orderly system, the heavens
and the Earth – both the product of a guiding
mind. Different cultures have developed very
different versions of the cosmos, with
different explanations for the patterns of stars
in the night sky and their implications for
humans and how we should live.

Physicists and astronomers prefer the term
"universe", meaning all the matter and energy
we can observe. Cosmology is the study of
how the universe started and how it has
developed in the billions of years since the big
bang, with projections into the distant future.

**In 100 words**

The universe is all the matter and energy we can observe. Indeed, it's possible that it is infinite, extending far beyond what we can observe. The **cosmological principle** says that, on a large scale, the universe is homogeneous (matter is spread roughly uniformly) and isotropic (the same in all directions). In other words, there is nothing special about the bit of the universe we can see. An observer situated billions of light years away would see much the same as we do – stars, galaxies, dust clouds and so on. But no one knows if the principle is true!

**WHY IT MATTERS**
Suggests that we are in a typical point in the universe

**KEY THINKERS**
Isaac Newton (1643–1727)
Alexander Fridman (1888–1925)
Georges Lemaître (1894–1966)

**WHAT COMES NEXT**
Astronomers are examining evidence that suggests the principle is unreliable

**SEE ALSO**
The heliocentric model, p.145

# The cosmological principle

In some cultures, people see themselves as just a part of nature. In others, humans are at the centre of things, with nature provided for our benefit, to be exploited as we wish. Since the time of Copernicus, astronomers have recognised that our part of the universe is nothing special – the Earth orbits the Sun, the solar system is part of one galaxy among many billions.

"Not only is the universe stranger than we think, it is stranger than we *can* think."

**Werner Heisenberg**, quantum pioneer

# Hubble's law

In 1924 Edwin Hubble's observations of variable stars in distant cloudy regions of space (known at the time as "nebulae") showed that they were in fact far outside the Milky Way and were galaxies in their own right. He went on to classify many galaxies according to their shapes. Now we know that there are as many galaxies in the universe as there are stars in our own galaxy.

Hubble went on to show that distant galaxies are moving away from us. He did this by measuring the **redshift** of their light (see page 58). At the time, the explanation for this was very uncertain.

Hubble was unlucky not to win a Nobel Prize for his work. He was nominated by physicists on the committee, but this was not allowed because, at the time, astronomy was regarded as a separate science from physics; this strict rule was abandoned shortly after Hubble's death.

**Hubble's law** states that the greater the distance between two galaxies, the greater the **velocity** with which they are moving apart. Hubble deduced this from observations of 46 galaxies whose light showed increasing **redshift** with distance from Earth. Although his measurements of redshift and hence velocity were good, his measurements of the distances to far-off galaxies were serious underestimates, by a factor of about seven. Two years before Hubble published his findings, Georges Lemaître published a similar theoretical result, based on Einstein's ideas (see page 78). He proposed that the universe itself might be expanding.

# 89

# The Big Bang

The more distant a galaxy is from us here on Earth, the faster it is observed to be moving away from us (see page 160). If we imagine turning the clock back, all galaxies must have been closer together in the past. The conclusion is that, at some time in the past, all galaxies emerged from a single point. This is known as the Big Bang theory.

The motion of galaxies is deduced from the **redshift** of their light that reaches us. There can be two explanations for the redshift. One is that the galaxies are moving rapidly away from us, *through* space. The light waves they emit are stretched out as they leave the moving source (see page 58).

The alternative is to imagine that the galaxies are being carried apart by the expansion of space itself. After light waves have been emitted by a galaxy, they travel through space whose expansion stretches the waves, increasing their wavelength and hence pushing them towards the longer-wavelength red end of the spectrum. This version of the redshift is known as the cosmological redshift.

The Big Bang is the idea that all the matter and energy that now make up the universe originated in a single point roughly 13.7 billion years ago. Since then, the universe has been expanding.

In its early stages, the universe was exceedingly hot, but as it expanded it cooled. Its temperature today is just 2.7 degrees above **absolute zero**. Matter condensed out to form the earliest stars and galaxies, held together by gravity. Light waves from that early time have since been stretched by the expansion of space. What remains are long-wavelength radio waves, the **cosmic microwave background radiation**.

# 90

# Nucleosynthesis

Astronomers are fond of saying that "we're all made of stardust". By this, they mean that many of the atoms of which our bodies are made originated in stars and spread out into space when these stars reached the end of their lives (see page 157).

Only the lightest **elements** formed after the Big Bang – hydrogen, helium and lithium. These elements are still the commonest in the universe. Heavier elements formed by **nuclear fusion** of these light elements in stars, which went on to explode as **supernovae**. The heaviest elements are thought to have been formed during energetic collisions between neutron stars.

The solar system formed 4.5 billion years ago from dust and gas containing all these elements. That's the "stardust" to which astronomers are referring. Since many of these elements are essential for life, it follows that no Earth-like life could have existed in the early years of the universe.

Shortly after the Big Bang, the particles present were moving too rapidly to fuse but, as the universe expanded and cooled, **quarks** bound together to form protons and neutrons that then formed nuclei of the lightest **elements**. This phase of **nucleosynthesis** (nucleus-forming) lasted until just 20 minutes after the Big Bang. **Nuclear fusion** in stars, in **supernovae** and in collisions between neutron stars produced heavier elements - all the elements in the periodic table up to element 94 plutonium (see page 108). Heavier elements have been produced artificially but these are mostly very unstable and decay rapidly.

# Dark matter

To stay in orbit around the Sun, each planet must travel at just the correct speed. Too slow, and the Sun's gravitational pull will cause it to spiral inwards; too fast, and it will disappear out into the depths of space. The planets furthest from the Sun travel most slowly around their orbits. That's because the Sun's gravity is weakest out there.

You might think the same result would hold for galaxies: the stars orbiting furthest from the galactic centre would travel most slowly. But that's not what Vera Rubin found when she measured the speeds of stars in spiral galaxies. Her findings showed that the outermost stars were the fastest, suggesting that they were experiencing a stronger gravitational pull than the inner stars. Her conclusion was that there must be some extra matter spread throughout the galaxy, increasing the galaxy's gravitational pull on the outermost stars. This matter that we cannot see is known as dark matter.

Dark matter is needed to account for the motion of stars as they travel around their galactic centres. It is called "dark" because astronomers cannot see it with any of their telescopes. They think it interacts only through **gravitational attraction** (see page 72), suggesting that it has **mass** and hence is a form of matter. Estimates suggest that there is five times as much dark matter as ordinary matter in the universe. Astronomers are not certain what particles make up dark matter. The particles may be different from those of the **standard model** (see page 123).

"As the Earth moves around the Sun and the Sun moves around the galactic centre, through the dark matter halo of our galaxy, a wind of dark matter particles appears to blow at us from the direction of Cygnus. Measuring the direction atoms travel after they have been struck by dark matter particles in this wind is potentially a powerful discriminator between a dark matter signal and terrestrial backgrounds."

**Jocelyn Monroe**, astronomer

# Dark energy

As the universe expands outwards from the Big Bang, it seems reasonable to assume that gravity will slow down the expansion and eventually the universe will collapse back inwards in a "big crunch". However, observations tell us that this is not what is happening.

Looking far into space means looking back in time. The most distant stars we see are the farthest back – close in time to the Big Bang. By measuring the **redshifts** of stars and **supernovae** at different points in the history of the universe we can see that the expansion of the universe is speeding up. Astronomers propose dark energy as the mechanism for this.

Einstein's mass-energy equivalence relationship tells us that we can think of this mysterious energy as a form of mass, spread thinly throughout space (see page 102). However, it is still very uncertain what exactly this stuff is.

Physicists propose dark energy as the mechanism driving the expansion of the universe. Their observational evidence for this comes from measurements of the velocities of **supernovae** at different times in the past.

The nature of dark energy is uncertain. Astronomers think it is spread uniformly throughout space and, although its **density** is very low, it still accounts for about 68% of the density of the universe, with 27% dark matter and just 5% conventional matter.

An alternative explanation for the **acceleration** is that the **gravitational force**, normally an attractive force, might have a repulsive effect at the greatest distances.

# 93

# Time travel

Journeying to the past or to the future is the stuff of science fiction. H G Wells' novella *The Time Machine* popularised the idea of inventing a device that allowed the user to move back and forth in time. Its protagonist, the Time Traveller, visits societies of the future and even sees the gradually reddening Sun as it approaches the end of its life (see page 157).

Looking out into space we are looking back in time because the light from the stars left its source some years ago. But no one has yet found a way to visit a specific point in the past; nor have we any convincing evidence of visitors or information reaching us from the future.

**WHY IT MATTERS**
Can we visit the past?

**KEY THINKERS**
Albert Einstein
(1879–1955)
Kurt Gödel (1906–1978)
Stephen Hawking
(1942–2018)
Kip Thorne (1940–)
Frank Tipler
(1947–)

**WHAT COMES NEXT**
The past might come
next!

**SEE ALSO**
Space–time, p.32

In
**100**
words

Time travel is the idea that we can move to a different point in time just as we can move through the three dimensions of space. One idea is that there might be a path through **space-time** (see page 32) that would lead us back to a point earlier than the time at which we started.

A serious problem with time travel into the past is the "grandfather paradox". If you travelled back and murdered your grandfather before your father was conceived, you yourself could never have been born. This defies the basic physical idea of **causality**.

Physicists talk of "a theory of everything". Such a theory would unite existing ideas that explain the **elementary particles** and the forces between them, as described by the **standard model**, with a new theory of gravity. Existing theories of **quantum mechanics** and general relativity have been tested successfully many times in their respective fields but they have yet to be combined in a single unified theory.

One candidate theory is **string theory**, in which point-like particles are thought of as one-dimensional strings that can propagate through space-time. In such a theory, **space-time** has many more dimensions than the familiar four.

## Theories of everything

One of theoretical physicists' objectives is to join up theories from different parts of physics. It would be no good having one theory that explained the history of the universe and another that explained the forces that act between particles if the two theories were contradictory.

In the 19th-century physicists showed that electricity and magnetism were aspects of the same force and that light was a form of **electromagnetic wave** (see page 68). In the 20th century, **quantum mechanics** showed how three of the four fundamental forces – electromagnetism and the two **nuclear forces** – could be united with the particles of the **standard model** (see page 123). But that leaves gravitation as a separate force, currently explained by Einstein's general theory of relativity (see page 78). The hunt is on for a theory of quantum gravity.

**WHY IT MATTERS**
Looking for ways of uniting separate theories

**KEY THINKERS**
Kurt Gödel (1906–1978)
Stephen Hawking (1942–2018)
Edward Witten (1951–)
Fotini Markopoulou-Kalamara (1971–)

**WHAT COMES NEXT**
Who knows? (But Kurt Gödel said such a theory was impossible)

**SEE ALSO**
Physics theories, p.176

# Doing Physics

This book looks at 100 key ideas in physics. But physics isn't simply a collection of ideas; it's also a way of finding out about the world, from the largest astronomical scale to the microscopic world of the tiniest particles. The ideas of physics are a result of centuries of thinking, observing and experimenting. The theories physicists have developed over the years have been remarkably successful in explaining these observations. But many loose ends remain – and some loose beginnings, too. There's always more physics to be explored, to be argued over and to be understood.

Observation is at the heart of physics as we try to gather information about Nature. Observers must take care not to allow their preconceptions to determine what they see. In physics, a single observation is rarely conclusive. Scientists must make multiple observations and other physicists must show that they are reproducible. This is why the publication of detailed results is essential.
While we may say that "seeing is believing", physicists have gone far beyond using just their eyes. Scientific instrumentation (using cameras, telescopes, detectors and meters of all types) allows many invisible phenomena to be observed, recorded and analysed

**WHY IT MATTERS**
Observation is at the start of understanding natural phenomena

**KEY THINKERS**
Francis Bacon
(1561–1626)
Robert Hooke
(1635–1703)
Carl Gustav Hempel
(1905–1997)

**WHAT COMES NEXT**
Careful observation allows the development of reliable scientific theories

**SEE ALSO**
Measurement, p.174
Experimentation, p,178

# Observation

In 1903, Prosper-René Blondlot announced the discovery of a type of radiation he called N-rays. He and others saw these rays coming from many different objects when observed in a darkened room. Unfortunately for him, many other scientists failed to reproduce his observations and it was soon shown that the "brightening" of the objects Blondlot studied was a physiological effect in the eyes of the observer. By that time, however, more than 100 scientific papers had been published on the subject.

The story of N-rays is a cautionary tale. Blondlot was working shortly after the discovery of X-rays, so the idea of invisible rays was in the air. Other physicists showed an insufficiently sceptical approach when repeating his work. His ideas were destroyed when a sceptical physicist removed a "source" of N-rays, but Blondlot continued to claim that he could see them.

# 96

# Measurement

In 1999, NASA's Mars Climate Orbiter was lost as it was about to enter its orbit around the planet. NASA attributed this failure to a mismatch between two pieces of software controlling the thrust produced by the rocket motors. One provided values in pound-force seconds (US customary units) while the other was programmed to expect values in newton-seconds (metric units). The difference was a factor of 4.45, which caused the spacecraft either to burn up or to fly off into orbit around the Sun.

Scientists, engineers, manufacturers and many others depend on reliable measurements, and each measurement must have a unit. Physicists generally work in units of the **SI (Système Internationale)**. Physicists around the world require measurements of natural phenomena to be increasingly precise, and they have redefined these units so that they can check that their instruments are correctly calibrated.

To check that their measuring instruments are giving reliable values, physicists must ensure that they are correctly calibrated. They also need to know how precise their measurements are since no measurement can be exact. Any uncertainty in a measurement is shown with a plus-or-minus sign (e.g. 10.27 ± 0.03 g).

Physicists use the metric **SI system** of units. This is based on fixed values of certain physical constants such as the speed of light in a vacuum and the charge of an **electron**. All other units are derived from these. A shared system of units allows comparison of results between laboratories.

| BASE QUANTITY | | BASE UNIT | |
|---|---|---|---|
| NAME | SYMBOL | NAME | SYMBOL |
| Length | l, h, r | meter | m |
| Mass | m | kilogram | kg |
| Time | t | second | s |
| Electric current | I, i | ampere | A |
| Temperature | T | kelvin | K |
| Amount of substance | n | mole | mol |
| Luminous intensity | Iv | candela | cd |

The seven base quantities and their units of the Système Internationale of scientific units

# Physics theories

The Big Bang theory, as well as being the name of a television show, is physicists' way of explaining how the universe came into being, developed and may end. Someone might say: "But it's only a theory", perhaps because it conflicts with religious teachings that they accept. However, this would be to misunderstand what a theory is in physics, and in science more generally.

The Big Bang theory (see page 162) is a set of ideas that have been tested over and over again. Several major pieces of evidence support the theory – the expansion of space, the **cosmic microwave background**, the proportions of the **elements**. The theory itself has been modified many times in the light of evidence so that physicists are increasingly confident that it is reliable. However, it would take only one contradictory experimental finding to cause a rethink.

In physics, a theory is not "only a theory"; rather, it's an idea with a body of supporting observational and experimental evidence. While physicists are always happy to find new evidence to support a theory, it is also important to devise experiments that could show that the theory is incorrect.

Albert Einstein suggested that there are two types of theory. One is a **model**, like the kinetic model that describes matter in terms of particles (see page 10). The other is a set of generalised ideas like Newton's theory of gravitation (see page 72).

# 98

# Experimentation

Physicists are often divided into experimentalists and theoreticians. However, each group needs to understand the work of the other because experiments must be guided by theory and theories must account for experimental evidence.

It's ironic that experiments in particle physics, searching for the smallest particles, require increasingly large experimental set-ups – think of the particle colliders at a lab such as CERN, for example (see page 70). These can be very expensive, requiring large numbers of engineers, technicians and scientists to build and operate them. The resulting scientific publications may list hundreds of authors, all of whom have made a contribution to the work and who deserve to get a share of the credit.

Experiments in physics produce new
data. The experimenter's motivation may
be to make new observations, to check previous
results or to test a theoretical idea.
Experiments must be carefully designed to produce
useful data. The outcome of an experiment may be determined
by a number of variable factors: each of these variables must
be carefully controlled, varying just one at a time, so that
the effect of each variable can be separately
determined. In some fields of study such as
astronomy it may be impossible to
set up practical experiments, so
data is gathered by
carefully controlled
observations.

# 99

# Mathematics

**WHY IT MATTERS**
The laws of physics are
mathematical

**KEY THINKERS**
Isaac Newton
(1643–1727)
Bernhard Riemann
(1826–1866)
Emmy Noether
(1882–1935)
John von Neumann
(1903–1957)

**WHAT COMES NEXT**
Artificial intelligence
may stimulate new
theories

**SEE ALSO**
The laws of
thermodynamics, p.98
$E = mc^2$, p.102

There are only a couple of equations in this book, because this is a book of key ideas. However, physics is intrinsically mathematical. When physicists publish their ideas, there are equations, tables, graphs and other bits of mathematics.

Sometimes physicists have had to invent mathematical techniques to explain their theories; Isaac Newton invented calculus for his laws of motion. At other times, the necessary mathematics has already been worked out by mathematicians (for whom it may have been an interesting problem without practical application). For example, Albert Einstein used Bernhard Riemann's work on multidimensional spaces in his work on **space-time**.

Galileo famously said that "the book of nature is written in mathematics". By this, he meant that the laws that govern physical phenomena are mathematical. They relate quantities such as **mass**, temperature and energy that can be measured or calculated. They can be written down as equations or represented using graphs or other mathematical diagrams.

Increasingly, computers are used to analyse large datasets, looking for underlying patterns that may hint at a new law. Supercomputers are also used to explore theoretical models which may relate many different variables. But no-one really understands why mathematics is "the language of nature".

"Mathematics began to seem too much like puzzle solving. Physics is puzzle solving, too, but of puzzles created by nature, not by the mind of man."

**Maria Goeppert Mayer,**
physics Nobel prize winner

# 100

# The unity of science

Biochemists are developing a new theory of how life originated on Earth, around deep ocean vents where chemical reactions could have acted as a source of energy before **photosynthesis** evolved. As well as drawing on knowledge of chemistry and biology, this theory requires a detailed understanding of the electric fields around atoms in order to predict how chemical reactions will proceed. The ideas of physics are increasingly merging with those of other branches of science: astronomy, chemistry, biology and so on.

**"We are just an advanced breed of monkeys on a minor planet of a very average star. But we can understand the universe. That makes us something very special."**

**Stephen Hawking**, theoretical physicist

Science, including physics, is one of humanity's greatest intellectual achievements. Ultimately, the ideas of physics must be self-consistent, and consistent with the ideas of the other sciences. Many of the most interesting puzzles in science are tackled by teams of scientists drawn from different disciplines.

Physics teaches us sceptical thinking. The ideas of physics are not set in stone; rather, they keep evolving. Some changes are revolutionary, others are gradual. But every theory is open to challenge, through observation and experimentation. There can only be one physics, across all nations and cultures, because the ultimate challenge is the evidence.

# Glossary

**Absolute zero:** The lowest temperature possible when all atomic movement has its minimum value. Absolute zero is 0 Kelvin, −273.15 Celsius and −459.67 Fahrenheit.

**Acceleration:** The rate of change of velocity, whether that is speeding up, slowing down, or changing direction. Acceleration is measured in m/s².

**Antimatter:** Matter made of antiparticles whose properties such as electric charge are the opposite of the particles which make up regular matter.

**Atomic number:** The number of protons in an atom's nucleus (also called the proton number). Each element has a different atomic number.

**Bonds:** An attractive force between atoms or molecules that holds them together, made or broken by chemical reactions.

**Causality:** The idea that one event, the cause, gives rise to another, the effect.

**Chemical potential energy:** Energy stored in a chemical substance that has the potential to be released during a chemical reaction, e.g. as heat and light.

**Conduction:** The movement of heat or electricity through a material.

**Conserved quantity:** Any quantity whose total value remains constant during an interaction; e.g. energy, momentum.

**Convection:** The transfer of heat through a fluid (a liquid or gas) caused by the tendency of hotter, less dense fluid to rise and cooler, more dense fluid to sink.

**Cosmic microwave background radiation:** Infrared radiation spread throughout the universe, left over from a time shortly after the big bang.

**Cosmological principle:** The idea that no particular region of the universe is very different from any other region.

**Crystalline:** A material whose atoms or molecules are arranged in a regular array.

**Density:** The mass of a substance per unit volume. Measured in kg/m³ or g/cm³.

**Diffraction:** The spreading out of a wave such as light or sound on passing through a narrow gap or around the edge of an object.

**Dispersion:** The separation of waves of different frequencies into a spectrum.

**Doppler effect:** The change in apparent frequency of a sound wave when the source and receiver of the sound are moving relative to one another.

**Electromagnetic spectrum:** All types of electromagnetic wave arranged according to their frequencies or wavelengths.

**Electromagnetic wave:** A periodically varying wave of electric and magnetic fields.

**Electron:** One of the three main particles in an atom (along with the proton and neutron).

**Elementary particle:** A particle such as the electron which is thought to have no underlying structure.

**Element:** A chemical substance whose atoms all have the same atomic number.

**Entropy:** A measure of the randomness in an arrangement of particles.

**Event horizon:** The boundary of a black hole; light within the event horizon cannot escape so that we can know nothing of what is inside.

**Free electrons:** Electrons in a metal which are not attached to particular atoms so that they can move within the bulk of the metal.

**Gauge boson:** An elementary particle that transmits a force between other elementary particles.

**Gravitational field:** A field of force caused by the mass of an object and which acts on any other object with mass.

**Gravitational force:** A fundamental force acting between any two objects with mass.

**Gravitational redshift:** The change in frequency of light as it travels out of the gravitational field of a massive object such as a star.

**Gravitational time dilation:** The slowing of time as observed by an object in a strong gravitational field.

**Heat capacity:** The energy required to raise the temperature of an object by 1 degree C.

**Higgs boson:** An elementary particle whose interaction with matter causes it to have mass.

**Hubble's law:** The law relating the speed of recession of galaxies with their distance from our own galaxy.

**Interference:** When two waves meet, they may combine to give a larger wave (constructive interference) or cancel out (destructive interference).

**Inertia:** The tendency of an object to remain stationary or to move with constant velocity, due to its mass.

**Ions:** atoms or groups of atoms which are charged because they have gained or lost electrons.

**Isotopes:** Atoms of an element which have different masses because their nuclei have different numbers of neutrons.

**Kinetic energy:** The energy of an object due to its motion.

**Leptons:** Elementary particles such as electrons and neutrinos and which, unlike quarks, do not experience the strong nuclear force.

**Magnetosphere:** The region around the Earth where its magnetic field interacts with charged particles arriving from space.

**Mass:** A fundamental property of matter which causes it to have inertia and which gives rise to the force of gravity.

**Model:** A simplified version of a physical situation, devised to explain observations and make predictions.

**Momentum:** The mass of an object multiplied by its velocity.

**Nuclear fission:** The formation of lighter atomic nuclei when a heavy nucleus splits apart.

**Nuclear force:** One of two fundamental forces, weak or strong, which act between particles in the atomic nucleus.

**Nuclear fusion:** The formation of a heavier atomic nucleus when two lighter nuclei join together.

**Nucleosynthesis:** In stars, the formation of atomic nuclei from smaller particles.

**Nucleus:** The positively charged central core of an atom, having most of the atom's mass.

**Photoelectric effect:** The release of electrons from a metallic surface when light is shone on it.

**Photosynthesis:** The chemical reaction in plants by which carbohydrates are formed from water and carbon dioxide.

**Positron:** The antimatter particle corresponding to an electron.

**Potential energy:** Energy of a particle due to its position in a force field.

**Quantum (plural – quanta):** The minimum amount of any physical quantity such that the quantity can only have certain discrete values.

**Quantised:** Having a limited range of possible values.

**Quantum mechanics:** The laws of physics obeyed by matter at the microscopic scale.

**Quarks:** Elementary particles which experience the strong nuclear force so that they combine to form particles such as protons and neutrons.

**Qubit:** The basic unit in quantum computing, equivalent to a binary bit in conventional computing.

**Redshift:** The decrease in the frequency of light when the source is receding from an observer.

**Refraction:** The bending of light as it passes from one medium to another, caused by its changing speed.

**SI system (Système Internationale):** The system of units used in science, based on fundamental units such as the metre, kilogram and second.

**Space-time:** A mathematical model which combines the three dimensions of space with the fourth dimension, time.

**Standard model:** the set of elementary particles which combine to form all of matter, together with the force between them.

**String theory:** A model of the underlying structure of nature in which elementary particles are replaced by one-dimensional strings.

**Supernova:** An exploding star as it approaches the end of its life thermal equilibrium: when there is no net transfer of energy between two objects in contact.

**Time-dilation:** The slowing of time when an object moves at speeds approaching the speed of light.

**Time's arrow:** The apparent direction of time, from past to present to future.

**Ultrasound:** Sound whose frequency is too high for human hearing to detect uniform motion: motion at a steady speed in a straight line.

**Velocity:** Speed of an object in a given direction.

**Wave-particle duality:** The idea that, to understand the behaviour of light, electrons etc, we may have to think of them as either waves or particles, depending on the situation we are considering.

# Index

# About the author

**David Sang** is the author of over 140 textbooks used in secondary schools around the world. He was a research physicist at Leeds University for nine years before turning to teaching. He has taught Physics in a high school, a sixth-form college and a university. Now he devotes himself to writing textbooks and developing other teaching resources, including multimedia. He has edited books and websites aimed at helping teachers to teach Physics, and was formerly the physics editor of *Catalyst* magazine.

## In Association with The Science Museum

The Science Museum is part of the Science Museum Group, the world's leading group of science museums that share a world-class collection providing an enduring record of scientific, technological and medical achievements from across the globe. Over the last century the Science Museum, the home of human ingenuity, has grown in scale and scope, inspiring visitors with exhibitions covering topics as diverse as robots, code-breaking, cosmonauts and superbugs. www.sciencemuseum.org.uk.

The publisher would like to thank the following for their kind permission to reproduce their photographs:

(Key: a-above; b-below/bottom; c-centre; f-far; l-left; r-right; t-top)

**1 Shutterstock.com:** T-flex. **2–3 Depositphotos Inc:** image4stock. **4–5 Shutterstock.com:** Taawon Graphics. **8 123RF.com:** Watchara Khamphonsaeng. **13 Depositphotos Inc:** newb1. **14–15 Shutterstock.com:** Free Ukraine and Belarus. **17 123RF.com:** Maria Gorbacheva (br). **36 Shutterstock.com:** etcberry. **39 Shutterstock.com:** Alexandr III. **40–41 Shutterstock.com:** Julee Ashmead. **44 Shutterstock.com:** suns07butterfly. **48–49 Shutterstock.com:** Morphart Creation (b). **52–53 Shutterstock.com:** Distance0. **56–57 Alamy Stock Photo:** Photo Researchers / Science History Images. **58 Shutterstock.com:** Alex_Murphy (br). **60 Shutterstock.com:** Morphart Creation. **65 Science Photo Library:** (b). **66–67 Shutterstock.com:** delcarmat. **74–75 123RF.com:** garrykillian. **76–77 123RF.com:** udaix. **79 Depositphotos Inc:** vectorguy. **82–83 123RF.com:** sergeiminsk (b). **86–87 Shutterstock.com:** Curly Pat. **89 Depositphotos Inc:** cmeree. **94–95 Shutterstock.com:** aunaauna. **96–97 Depositphotos Inc:** image4stock. **102–103 Shutterstock.com:** Mykola Mazuryk. **104 Shutterstock.com:** Olgastocker. **108–109 123RF.com:** Humdan Maseng. **116–117 Shutterstock.com:** End Art. **118–119 123RF.com:** Denis Barbulat. **122–123 123RF.com:** sumkinn. **125 Shutterstock.com:** newelle. **128–129 Shutterstock.com:** Graphicindo42. **134–135 Shutterstock.com:** Alina_Bukhtii. **139 Shutterstock.com:** Arvila (br). **144 Shutterstock.com:** whitehoune. **147 Shutterstock.com:** ekosuwandono. **148–149 Shutterstock.com:** IfH (b). **155 Shutterstock.com:** ya_blue_ko (t). **157 Shutterstock.com:** Distance0. **158 123RF.com:** Aleksandra Alekseeva. **162–163 Shutterstock.com:** T-flex. **172 Shutterstock.com:** Fatima AIT ADDI. **180–181 123RF.com:** Sergii Pal

Cover images:
All other images © Dorling Kindersley

**DK LONDON**

**Editor** Florence Ward
**Art Editor** Anna Formanek
**Managing Editor** Pete Jorgensen
**Managing Art Editor** Jo Connor
**Production Editor** Jennifer Murray
**Production Controller** Louise Minihane
**Publishing Director** Mark Searle

**Written by** David Sang
**Designer** Neal Cobourne
**Jacket Designer** Steven Marsden

DK would like to thank Charles Phillips for copyediting, Alisa Jordan Walker and Caroline Curtis for proofreading, Vanessa Bird for indexing and Izzy Merry for design assistance.

First American Edition, 2024
Published in the United States by DK Publishing
1745 Broadway, 20th Floor, New York, NY 10019

A catalog record for this book
is available from the Library of Congress.
ISBN 978-0-7440-8162-6

DK books are available at special discounts when purchased in bulk for sales promotions, premiums, fund-raising, or educational use. For details, contact: DK Publishing Special Markets,
1745 Broadway, 20th Floor, New York, NY 10019
SpecialSales@dk.com

Printed and bound in China

**www.dk.com**

This book was made with Forest Stewardship Council™ certified paper – one small step in DK's commitment to a sustainable future.
**For more information go to www.dk.com/our-green-pledge**